Plantations in the Tropics

Environmental Concerns

IUCN – THE WORLD CONSERVATION UNION

IUCN – The World Conservation Union brings together States, government agencies and a diverse range of non-governmental organizations in a unique world partnership: some 720 members in all, spread across 118 countries.

As a union, IUCN exists to serve its members – to represent their views on the world stage and to provide them with the concepts, strategies and technical support they need to achieve their goals. Through its six Commissions, IUCN draws together over 5000 expert volunteers in project teams and action groups. A central secretariat coordinates the IUCN Programme and leads initiatives on the conservation and sustainable use of the world's biological diversity and the management of habitats and natural resources, as well as providing a range of services. The Union has helped many countries to prepare National Conservation Strategies, and demonstrates the application of its knowledge through the field projects it supervises. Operations are increasingly decentralised and are carried forward by an expanding network of regional and country offices, located principally in developing countries.

IUCN – The World Conservation Union seeks above all to work with its members to achieve development that is sustainable and that provides a lasting improvement in the quality of life for people all over the world.

THE IUCN COMMISSION ON ECOLOGY

The IUCN Commission on Ecology (COE) serves as the Union's source of technical advice for translating knowledge of ecological processes into practical action for conservation, sustainable management and restoration, in particular of areas degraded by human action. The IUCN programmes on Forest Conservation, Marine Conservation, and Wetlands are under the umbrella of the Commission on Ecology. Commission members serve on advisory committees and in working groups associated with these programmes.

THE IUCN FOREST CONSERVATION PROGRAMME

The IUCN Forest Conservation Programme (formerly the IUCN Tropical Forest Programme) coordinates and reinforces activities of the IUCN members and Secretariat which deal with forests. The Programme focuses on the conservation of species and ecological processes, and on investigating and promoting sustainable use of the resources of these forests.

The Programme includes international and national policy initiatives and strategies as well as field projects addressing selected problems in managing the world's most biologically significant tropical forests. These selected projects put the World Conservation Strategy into action by reconciling the requirements of conservation with national development and the needs of people living in forest areas. Special emphasis is given to the development of compatible uses for buffer zones around national parks and reserves.

IUCN develops its positions and policies on the basis of the concerns and information communicated by members, trends identified by monitoring activities, and the feedback from numerous field projects. Data on species of plants and animals, and on forest sites which are important for biological and ecosystem conservation, are held by the World Conservation Monitoring Centre in Cambridge, UK.

This series of publications from the Forest Conservation Programme, in conjunction with regular meetings, enables IUCN to communicate policies and technical guidance to governments, major international institutions, development planners, and conservation professionals. The Programme works closely with development assistance agencies, governments and NGOs, to ensure that conservation priorities are adequately addressed in their activities.

The Forest Conservation Programme receives generous financial support from the Government of Sweden.

The IUCN Forest Conservation Programme

Plantations in the Tropics

Environmental Concerns

Jacqueline Sawyer

IUCN / UNEP / WWF
1993

Published by: IUCN, Gland, Switzerland and Cambridge, UK in collaboration with UNEP and WWF

IUCN The World Conservation Union   UNEP   WWF

Citation: Sawyer, J. (1993). *Plantations in the Tropics: Environmental Concerns.* IUCN, Gland, Switzerland and Cambridge, UK. xiii + 83 pp.

ISBN: 2-8317-0139-2

Printed by: Unwin Brothers Ltd, Old Woking, Surrey, UK.

Cover design by: James Butler

Cover photo: *Eucalyptus camaldulensis* plantation in the South Iberian Peninsula: Manuel Ruiz Pérez

Produced by: IUCN Publications Services Unit, Cambridge, UK, on desktop publishing equipment purchased through a gift from Mrs Julia Ward

Available from: IUCN Publications Services Unit
181a Huntingdon Road, Cambridge, CB3 0DJ, UK
or
IUCN Communications and Corporate Relations Division
Rue Mauverney 28, CH-1196 Gland, Switzerland

The text of this book is printed on Reprise Offset 80 gsm recycled paper

CONTENTS

FOREWORD

Tropical deforestation has reached a historical maximum during the last decade and attracted considerable public attention. It can now be said that the extent of tropical deforestation and the problems associated with it are widely agreed upon. However, proposed solutions have not met with the same degree of consensus. Put simply, the debate has become divided between those who favour increasing the efforts made to curb deforestation and those who argue that deforestation (and greenhouse gas emissions) could be counterbalanced by increasing the extent of plantation cover.

Of course, if carefully implemented, both approaches would complement one another. Recently though, attention has focused on multipurpose plantations that not only supply wood, but also reduce pressure on surrounding natural forests, help restore degraded environments and fix carbon dioxide. And indeed, some tree plantations in the tropics have secured multiple benefits. Others, however, have created serious environmental and socioeconomic problems, or even failed altogether.

Concern about the environmental effects, particularly of large-scale plantations, prompted UNEP to initiate a study to clarify some of the issues involved. The resultant report, undertaken by IUCN, was also sponsored by WWF International. It seeks to illustrate both the potential of plantations and the problems that might arise if the massive tree plantations currently being suggested for the tropics were to be established.

The report is based on extensive bibliographical research into what has become a highly controversial topic, one which is in fact riddled with conflicting 'evidence', as was illustrated by the review process. During this process many recommendations were made which improved the first draft, but in addition a substantial number of contradictory comments were received from experts with similar backgrounds. The report has tried to integrate the comments received as far as possible.

The main body of the report is structured in terms of the major issues concerning large-scale plantations. The first issue covered is species selection. This is discussed in relation to plantation objectives and environmental site conditions. The tendency to concentrate on a few genera means that improved plants and technology are available for a limited number of species only. This has sometimes led to plantation failure due to mismatches between species and sites.

Admittedly some plantations provide good soil coverage, prevent erosion and help regulate the water cycle. But others, due to inappropriate practices when planting and harvesting, or during the construction of forest roads, have triggered serious erosion processes. Moreover, local water cycles have been disrupted as a result of periodic clearcutting, and in some cases, by increased rates of water uptake due the physiology of the species chosen and/or the density of planting.

Pests, diseases and other factors such as large fires have frequently been associated with large-scale plantations, and consequently attributed with many negative environmental and economic effects. Integrated pest management, genetic improvement, careful selection of species and adoption of best silvicultural techniques may help reduce risks and damage substantially. Unfortunately, lack of access to technology, together with the high costs incurred when applying risk prevention measures, mean that best practice is very rarely pursued.

The detractors of plantations have argued that, in comparison to natural forests, plantations reduce biodiversity, and much evidence has been gathered to support this claim. Nevertheless, it should be pointed out that plantations in the tropics frequently have higher biodiversity values than the agricultural fields surrounding them and that in some old non-harvested plantations biodiversity may approach that of a natural forest. For the sake of clarity then, large-scale plantations should be treated as what in fact they generally are: namely, long-rotation crops for the production of wood.

The recent 'discovery' that plantations represent a potential means of solving the climate change problem, due to their ability to store CO_2, is very controversial and has generated much confusion. Part of the confusion is due to the fact that most of the scenarios proposed use unrealistic growth rates, and figures relating to maximum standing biomass (rather than averages across the whole rotation cycle). Thus estimates referring to the quantity of carbon fixed are overly optimistic. Besides, in terms of overall forest cover, it is likely that the extent of future plantations in the tropics will be relatively insignificant and able to make only a small—though by no means negligible—contribution to efforts to combat climate change. Consequently, CO_2 fixation should not be cited as the main justification for establishing large-scale plantations, but seen instead as a complementary side effect.

The report also analyses various land tenure and social issues linked to large-scale plantations. Some plantations have been established in a top-down fashion, without local people's involvement and in opposition to the interests of communities with traditional rights over the planted land. However, other plantations have provided regular employment and income for rural populations, and when linked to a suitable processing industry, for urban populations as well.

Finally, some consideration is given to the economics of plantations, one of the areas of debate where most distortion has occurred. The frequent use of a purely financial (ie monetary returns on invested capital) versus a more comprehensive economic approach has precluded reaching a clear understanding of the costs and benefits of plantations. Emphasis on the latter might result in changes to the objectives of plantations, the techniques used to establish and manage them, and perhaps even to their size.

One of the report's main conclusions is that problems, when they do appear, tend to be linked not to the plantations *per se* but to the ways in which they have been designed and managed. IUCN therefore believes that drawing attention to the risks of poor planning and management practices may contribute to the improvement of existing and future plantations. A direct recommendation is that in-depth environmental and socioeconomic assessments should be carried out prior to planting.

A corollary of this is that evaluation of large-scale plantations has to be made in concrete terms, that is, on a case by case basis. So the final section of the report consists of only a few broad, general conclusions and some simple guidelines. And these should be seen in relation to what the IUCN Forest Conservation Programme considers the suitable framework: that of stopping deforestation. It is within this framework that plantations, properly managed, can play a positive role.

Manuel Ruiz Pérez
Forest Conservation Programme, IUCN

February 1993

ACKNOWLEDGEMENTS

Thanks are due to Julio Cesar Centeno (University of Los Andes, Venezuela), John Erskine (Institute of Natural Resources, University of Natal, South Africa), Julian Evans (Forestry Commission, UK), Philip Fearnside (National Research Institute of Amazonia, Brazil), Pieter van Ginneken (AIDEnvironment, the Netherlands), Helen Groome (University of the Basque Country, Spain), Peter Kanowksi (Oxford Forestry Institute, UK), Simon Rietbergen (World Bank), Manuel Ruiz (IUCN) and Peter Savill (Oxford Forestry Institute, UK), for their additions to the manuscript. Special thanks are due to the Forest Resources Division of FAO for its significant input. Any errors, however, remain the responsibility of the author.

GLOSSARY OF ACRONYMS AND ABBREVIATIONS

CTFT	Centre technique forestier tropical
EC	European Community
FAO	UN Food and Agriculture Organization
IPM	integrated pest management
IUCN	World Conservation Union
LDC	less developed country
UNDP	United Nations Development Programme
UNEP	United Nations Environment Programme
WWF	World Wide Fund For Nature
Gt	gigaton
ha	hectare
km	kilometre
mm	millimetre
MW	megawatt
Pg	petagram (the equivalent of 10^{15} grams or 10^{9} metric tons)
pH (value)	a number used to express degree of acidity or alkalinity in solutions
Tg	teragram (the equivalent of 10^{6} tons)
Tm	metric ton
yr	year

1. INTRODUCTION

Recent suggestions to establish plantation forests over hundreds of millions of hectares, especially in the tropics, as a means of combating climate change, have engendered much debate concerning the merits and demerits of this particular land use. IUCN therefore perceived a need for an overview of plantations, including the roles they fulfil, or fail to fulfil, and the problems associated with them. Care especially needs to be taken to distinguish between different types of plantation forestry, and to ensure that practices associated with poor forestry generally, are not used to typify large-scale plantation forestry.

This document has been written with the three main aims of the World Conservation Strategy in view (IUCN/UNEP/WWF, 1980), but does not attempt to be a complete guide or manual to any of the technical aspects of plantations. Many exist already, and it is these which should be consulted when in-depth discussion concerning the establishment and tending of plantations is required. Similarly, there are several in-depth environmental reviews of large-scale plantations (see, for example, Barnett, 1992), and it has not been thought necessary or helpful to reproduce something of that nature here. Instead, this document hopes to bring together for consideration some of the main issues surrounding what is fast becoming a confused and polarized debate.

2. PLANTATIONS—A DEFINITION

Plantations may be loosely defined as arrangements of trees which are artificially generated and the result of conscious management (Sedjo, 1987). Natural forests may also be consciously managed and artificially regenerated, although in parts only, but they will not have been completely created by man, or at least not recently. The idea of "a continuum of intervention", as depicted in Table 1, helps to illustrate this point.

Table 1. The continuum of intervention

Status	Level of Intervention
undisturbed natural forest	no intervention
protected natural forest	for example, control of insect pests or control of activities in adjacent areas
managed natural forest	activities to promote natural regeneration, such as enrichment planting, line planting
plantation on recently forested area	planting with 'new' material, but also retaining some relict trees
tree crop/plantation consisting entirely of planted trees	intervention at all levels, from establishment through to harvesting (of timber or non-timber products)

In size, plantations may range from under a hectare, as when trees are planted by a farmer as a windbreak, to several thousand hectares, as when a plantation is established to produce pulpwood for a paper mill. It is therefore difficult to make meaningful statements which can be applied to plantations in general.

2.1 Distinguishing Between Types

Various distinctions between plantations can be drawn. Some of the most common are: simple/complex; small-scale/large-scale; a plantation consisting solely of a forestry activity/a plantation comprising part of a broader activity. The objectives for which it is established determine which of the above categories describes a plantation. A simple plantation could be one created solely for the purpose of industrial production or for a single service such as watershed rehabilitation or protection. Simple plantations are often established over extensive areas and characterized by their ordered, regular appearance and relative ecological simplicity. Tree plantations established purely for food crops, for example mango or citrus plantations, would also be simple plantations, but they are usually considered as comprising an agricultural rather than a forestry activity and as such will not be dealt with here.

A complex plantation might be one designed to maximize social benefits rather than wood production *per se*, and which included elements of forestry activities such as agroforestry and community forestry (Kanowski and Savill, 1992). A complex plantation 'forest' may also cover a considerable area, but consist of many small-scale, individual units, rather than a single homogenous unit, and present a more irregular appearance than an extensive simple plantation.

Following Evans (1992) we can equate "small-scale" with plantations which have no centralized management and which singly are not large enough to meet the total industrial wood requirement of a commercial mill. Under this heading he includes village woodlots, plantations arising from extension work and social forestry programmes, farm plantings for shelter, fodder, and so on, trial plantings to evaluate species on different sites, scattered plantings arising from regional or national campaigns, and use of multipurpose trees in rural development and planting to contain localized or incipient environmental problems. "Large-scale" can then be used to refer to many of the industrial production plantations established to supply single enterprises such as pulp mills.

Plantations comprising a forestry activity in themselves can be understood as those which, to a certain extent, function independently of their surroundings and which are traditionally managed by foresters. Plantations established under a broader activity would include those forming part of the plans for the rural development of an area. They are therefore more likely to involve disciplines other than forestry, such as sociology and anthropology.

A distinction that Groome (1991) uses in relation to forestry generally, can be applied to plantations in particular. She talks of horizontal and vertical integration and as an example of the former describes a woodland which meets several needs at the same time. Vertical integration is illustrated by the example of a woodland which supplies a factory which in turn supplies trade. A multipurpose agroforestry plantation would illustrate horizontal integration and an industrial timber plantation, supplying a pulp mill, would illustrate vertical integration.

Tentatively, then, we can identify two extremes:

simple	complex
large-scale	small-scale
consisting solely of a forestry activity	comprising part of a broader activity
vertically integrated	horizontally integrated.

The qualities listed on the right-hand side might reasonably be expected to apply to those plantations which are most integrated with their surroundings. However, many plantations will fall somewhere between the two extremes. In the United Kingdom, for instance, the Forestry Commission has established large-scale plantations for timber production, but which also provide recreation facilities and which are planned and now restructured to harmonize with the local landscape. Moreover, even a large-scale, high technology plantation will develop an understorey and herb layer, adding some complexity to its structure and species composition. Likewise, in many developing countries, small-scale planting often forms part of a broader activity such as stabilization of agriculture, or regional development, but may simply consist of shelterbelts or woodlots of single species. But no matter which definition is favoured, all plantations have social, economic and environmental effects.

Figure 1: Horizontal and vertical integration

Source: Groome (1991)

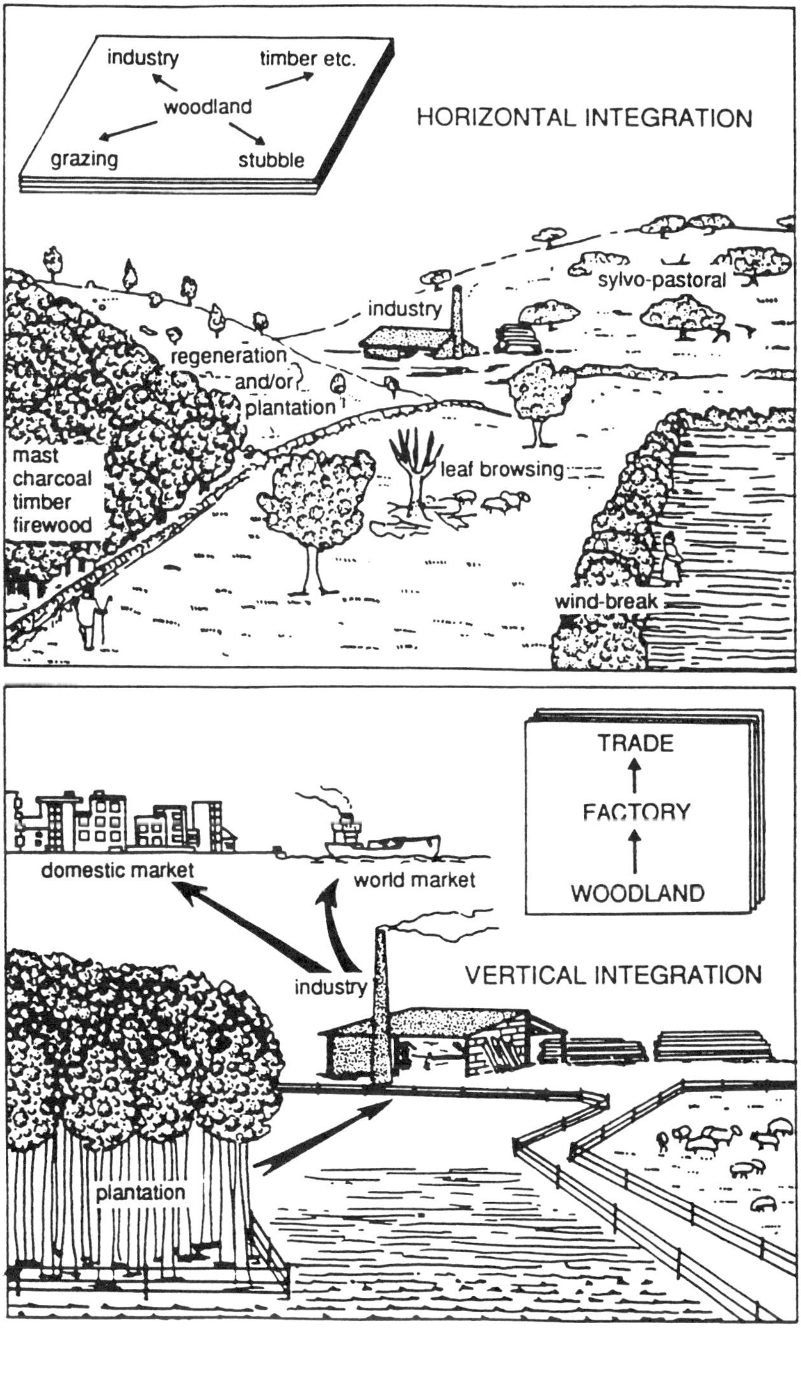

2.2 Management Strategies and Technology

In the last 50 years or so management strategies for simple plantation forestry, particularly for large-scale timber production, have become very advanced technologically, although they are not always applied satisfactorily. Those for complex plantation forestry are not yet technologically advanced (Kanowski and Savill, 1992), perhaps because they favour a diversified rather than concentrated approach. An additional factor is that funding for the development of plantation technology often comes from the private sector which is interested mostly in production forestry. And of course, the success of complex plantation forestry is not so dependent on (high) technology anyway.

Technology is also usually correlated with size. It may be worthwhile to use sophisticated machinery to prepare large areas all at once for planting a single species, for a single purpose. But it would not be worthwhile to do so over fifty smallholder plots, each with its own distinct boundary, and on several separate occasions. Something of a paradox in fact operates whereby high technology is usually associated with large-scale plantations which represent relatively simple systems, while low technology often characterizes small-scale plantations which nevertheless represent sophisticated systems (see also Section 7). An extreme example of the latter would be a home garden in Java which brings together many tree, shrub and ground crop species and which is managed in an integrated way.

This is not to say that some of the technologies used in simple plantation forestry could not also be applied more often in complex plantation forestry. On the contrary, if careful adaptations were made, complex plantation forestry could benefit from technologies such as genetic improvement which are currently used to enhance the health and productivity of timber production plantations.

2.3 Confusion and Conflict

Perhaps not surprisingly the debate surrounding plantations has become rather muddied. Criticisms or justifications are frequently made without adequate reference to the type of plantation being referred to or to its objectives. Additionally, research applying to one particular site or site type is sometimes extrapolated to several sites or site types, leading to inaccurate generalizations. For example, it has been claimed that conifers and *Eucalyptus* species decrease soil pH. This may be true in cases where conifers have been planted on a site with a low buffering capacity, but would not be true if the site had an alkaline soil with a high buffering capacity, or if the soil was already very acidic.

Research results concerning plantations are also on occasion presented misleadingly. Research may be funded by an organization or group with a particular interest, but this may not be made clear when the results are presented. Alternatively, the interest is allowed to bias the direction the research takes or the interpretation of results.

Ideally, given the vast array of biological, social and economic factors likely to be operating, each site or plantation would be evaluated individually before final judgement was made. Similarly, it cannot be stressed enough that each site will have different requirements in terms of appropriate species, silvicultural operations and conservation management.

3. OBJECTIVES OF PLANTATIONS

To date, large-scale plantations for timber production have dominated plantation establishment. But other uses are also described here in order to illustrate the potential of plantations for meeting demands for various products and services. Although these uses or objectives are listed here individually, there is increasing emphasis on plantations which combine several of them.

Unfortunately, theory and reality do not always coincide. Some plantations are established for valid reasons, but fail to attain their goal because of inadequate management. Timber plantations, for example, have been devastated by pests and diseases which were not identified quickly enough, and plantations for non-timber products have been destroyed by uncontrolled grazing. Others are planted for inappropriate reasons—for example for timber production when local markets are not well developed and exporting problematic. Yet others are planted ostensibly for one reason, such as watershed management, but actually for quite another, such as meeting a forest department planting target set by government.

And although there has been much discussion of sustainability in recent years, little consideration has been given to applying the concept in its broadest sense (see for example, Barbier, 1987) to plantations. Thought must be given to evaluating plantations in such a way that all their impacts are taken into account, be these environmental, social or economic. For example, although a plantation may be 'successful' in that it enables a country to reduce its imports of timber products, it may also be distinctly 'unsuccessful' since it affects soil and water quality negatively. To take another instance, a woodlot established on communal grazing land may be a vital new source of fuelwood, but marginalize the worst-off sectors of society even further.

It is therefore important that when planning plantations, objectives are realistic and defined clearly, and that management plans, if appropriate, reflect them adequately. Unless there is complete clarity from the very start about the purposes to be served by a plantation, it is likely that difficulty and controversy will arise as the enterprise proceeds (Poore, 1992). Obviously, not all objectives can be satisfied on each piece of land and some analysis of trade-offs may be required. Alternatively, if it is decided, for instance, that a plantation of fast-growing pioneer species to supply local industry must be established, it may be possible to offset or counterbalance this use by assigning priority to another use, such as wildlife conservation, elsewhere. But to be able to take such decisions, good land use planning is essential.

3.1 Timber Production

Many 'modern' plantations were established initially because of timber shortfalls. For example, the United Kingdom's forest cover had shrunk to 5% by the early 20th Century and when the First World War broke out, the strategic folly of not having reliable home timber supplies became clear. So in 1919 the Forestry Commission was set up, its main objective initially being to ensure a timber reserve (Forestry Commission, 1988). However, although the Commission's plantations have achieved their original target, they have been criticized on various grounds; for instance, many of them were planted on areas of conservation interest.

More recently, Malaysia's Compensatory Forest Plantation Programme was implemented because of fears that the country's natural forests would be unable to continue supplying raw material to its wood processing industries. The programme, which started in 1981, envisages the establishment of around 188,000ha of forest plantations in Peninsular Malaysia by 1995

(Baharudin, 1987). Commentators such as Binkley and Vincent (1992) have strongly criticized the programme, however. They claim that state and federal forestry departments have used it to justify high rates of harvest of the natural forest, and that this factor, plus rivalry between agricultural and forestry agencies, have motivated the programme, rather than sound economic objectives.

Elsewhere, in Nigeria, timber plantations of both native and indigenous species have been created because experiments with natural regeneration systems did not result in stocking or growth rates high enough to meet timber demand (Kio and Ekwebelam, 1987; Nwoboshi, 1987). However, it is possible that further experimentation with natural regeneration would have produced more encouraging results, as has been the case in Uganda and the Philippines, although regrettably not applied in these countries. Perhaps too the original results themselves should be reevaluated; many of the species which regenerated well under the Tropical Shelterwood System, such as *Piptadeniastrum* and *Combretodendron* spp., but which were not considered useful species then, are now marketable.

But timber production plantations are not established solely because existing, natural forests cannot meet demand. Natural forests may be of considerable extent, but located at such distance from markets and/or end users that felling and transporting costs would be unacceptably high. This is the case in Canada (Savill and Evans, 1986) and it is also one reason why the Amazon forest remains substantially intact (Evans, 1985). Alternatively, natural forests may be on swampy ground or steep, mountainous terrain and difficult or expensive to harvest (Evans, 1992). Plantations established near infrastructure can therefore be more cost effective and less wasteful of natural resources over the long term.

Another consideration is that natural forests may be providing important services or have non-timber values which override their timber production values. Montane forests protect against erosion and avalanches, for instance, while other forests have high conservation or amenity value (Evans, 1986a). By making it possible to designate natural forests as parks and reserves (Gladstone and Ledig, 1990) or to assign them to water catchment protection (Spears, 1983), plantations can help ensure that such services and values are maintained. Plantations may also act as buffers to natural forests (Fearnside, 1983).

Interestingly, it appears that greater recognition is being given to the non-timber values of natural forests. This is shown in part by the government forestry policies of around 30 developing countries, which favour increased dependence on plantation forestry as a primary source of industrial raw material, as opposed to an expansion of logging in natural forests (Spears, 1991).

Thus while plantations cannot provide the full range of goods and environmental services that are available from natural forests, they can contribute to their wise management; provided, of course, that natural forest lands are not then perceived as having no value, other than for clearance for agriculture (Ball, 1992).

Timber plantations are also established because they can be manipulated to produce much higher timber yields than natural forests, and quickly. Generally speaking, as rotations become longer or conditions more arid, mean annual increments become less. Productivity is highest in areas with seasonal rather than continuous rainfall, of up to 2,500mm annually. In this respect plantations parallel modern agriculture in their requirement of relatively high inputs and intensive management, near-maximum use of a site's productive capacity and high output of a narrow range of products. As knowledge of species and provenances increases, and experience in the use of exotics and application of genetic tree improvement accumulates, plantation growth rates can be expected to become even greater (Mather, 1990). However, genetic improvement also implies higher input costs, for example for fertilizers and pesticides, in order to maintain gains.

Another benefit of timber plantations is that they can be managed to produce a uniform product of specified qualities (which means less wastage) at a specified time. Aracruz Florestal's Programme in the state of Espirito Santo in Brazil has been particularly successful at developing plantations which produce timber of great uniformity, with higher average specific gravity and higher pulp yield per unit dry weight than wood from conventional forests (Campinhos Jr, 1991). This example could be multiplied many times, especially for pine, spruce and *Eucalyptus* plantations, and means that relatively small plantation areas can satisfy a society's industrial wood needs. This is an important factor in areas with a high population:land ratio, for domestic consumption and for export oriented crops.

Indeed, Sedjo (1987) has calculated that by the year 2000 plantations could supply 50% of total world wood supply, since by this time many of the plantations established in the 1960s will have reached maturity (see also Section 4.3). Previous calculations by Sedjo and Clawson (1984) show that, in theory, all of the world's industrial wood needs in the year 2000 could be supplied by between 100 and 200 million ha of well-managed plantations. In other words, we do not need unlimited plantations.

However, these calculations should be treated with caution. It might not be possible to guarantee the continuing productivity of Sedjo and Clawson's well-managed plantations and the siting of them could be the cause of some dispute were they to be planted. One is reminded too of the truism that present agricultural technologies and areas are sufficient to enable the world to easily feed its inhabitants, and the realities of politics, geography and history, which combine to preclude this.

Table 2. Increases in productivity of *Eucalyptus* plantations in Brazil

Period	Practices Adopted	Mean Productivity $m^3/ha/yr$
1960–1965	Low genetic quality seeds	12
1966–1970	Low genetic quality seeds and fertilizer usage	17
1970–1975	Genetically pure seeds (unimproved) and fertilizer usage	22
1975–1980	As in the previous period, from selected stand	35
1980–1985	Improved seeds and vegetative propagation	45
1985–1990	As in the previous period and additional selection	60

Source: Adapted from Campinhos Jr, 1991

Finally, it should be remembered that plantation-grown timber cannot always replace that from natural forests in terms of desired qualities. For example, many tropical hardwood species, some of which are used for specialist purposes, such as musical instrument or fine cabinet making, cannot yet be grown in plantations (see also Section 6.1). Additionally, as Sutton (1991) points out, it is doubtful that any but a few of the plantations now being established in the tropics will be managed for the production of quality sawlogs. They will rather be used as a source of supply of pulpwood or cheap construction timber. The only exception to this might be some teak plantations. It is therefore possible that a shortfall in quality tropical sawlogs will be experienced if natural tropical forests continue to be overexploited.

Table 3. Average growth rates attained in some plantations

Plantation Development	Species	Mean Annual Increment ($m^3ha^{-1}a^{-1}$)	Rotation (years)
Usutu forest, Swaziland	*Pinus patula*	19	15–17
Viphya Pulpwood Project Malawi	*Pinus patula*	18	16
Fiji Pine Commission	*Pinus caribaea*	15–20	17–20
Jarí Florestal, Brazil	*Pinus caribaea*	20	16
UAIC[1], Congo	*Eucalyptus* hybrids	35	7
Aracruz Florestal, Brazil	*Eucalyptus grandis*	55	7
Shiselweni Forestry, Swaziland	*Eucalyptus grandis*	18	9
PICOP[2], Philippines	*Paraserianthes falcataria*	28	10
Sabah Softwoods, Malaysia	*Paraserianthes falcataria*	30	7–10
Jarí Florestal, Brazil	*Gmelina arborea*	15–25	10
CNGT[3], Papua New Guinea	*Araucaria* spp.	20	40
Seaqaqa plantations, Fiji	*Swietenia macrophylla*	14	30

[1] Unité d'Afforestation Industrielle du Congo; [2] Paper Industries Corporation of the Philippines; [3] Commonwealth New Guinea Timbers

Source: Evans (1992)

3.2 Industrial and Regional Development

A fairly common use for plantations is as a basis for industrial and/or regional development. In several countries the creation of extensive plantations, mainly for wood production, has led to regional prosperity for some sectors of the population. Wood processing industries are set up in and around the plantations, community infrastructure developed and the economic base of the region expanded. This has occurred most notably in Australia and New Zealand, with large-scale *Pinus radiata* plantations. Similarly in the southeast of the United States, indigenous conifers (*P. elliottii* var. *elliottii* and *P. taeda*) have been planted on large areas of failed agricultural land, bringing prosperity to a formerly depressed area (Shepherd, 1986). Conversely, though, ill-planned plantations have led to reduced employment opportunities (see Section 13.6).

Reference is also often made to the fact that in several developing countries plantations have improved the balance of trade by reducing or eliminating the need to import industrial timber or wood products. Industrial plantations in Chile, Kenya and Zambia are good examples here (Spears, 1983). Some countries—including Australia, Brazil, Chile, New Zealand, Portugal, Spain and Swaziland—established plantations with the specific aim of exporting timber products (Gauthier, 1991). In fact, due to its large reforestation programme initiated in the 1960s, Brazil is no longer a net importer of pulp and paper but an important exporter (Suchek, 1991). However, costly government incentives were necessary to establish the plantations and neither their opportunity or environmental costs were fully analysed. This means that their true worth to the economy has perhaps been exaggerated. Of course, the same can be said of many large-scale plantations established elsewhere.

Table 4. Pulp and timber trade balance (in $US millions)—Brazil

	Paper			Pulp			
Year	Imports	Exports	Balance	Imports	Exports	Balance	Grand Total
1970	6	5	-1	48	1	-47	-48
1975	32	30	-2	116	9	-107	-109
1980	35	364	329	165	156	-9	320
1985	15	279	264	98	263	174	438
1990	42	598	556	244	603	359	915

Source: Suchek, 1991

3.3 Energy

Wood can provide energy in various forms; directly, in the form of fuelwood, or indirectly, as wood gas, ethanol, charcoal, etc, for either domestic or industrial use. In fact, one of the main reasons for social forestry in developing countries is to provide fuelwood for rural populations in areas where existing fuelwood supplies are overexploited and/or inadequate (see also Section 11.1).

Similarly, if energy demand is high for a given area such as a city or town, or for a particular industry, such as charcoal production or tobacco curing, establishing energy plantations can prove very useful (Davidson, 1987). At the turn of the century, plantations of *Eucalyptus globulus* were established around Addis Ababa because of fears that the recently established capital would be abandoned due to an increasing fuelwood shortage. By 1975 the net area of these plantations had reached 15,000ha (Pohjonen and Pukkala, 1990). And in Brazil, in the state of Minas Gerais, over 200,000ha of private commercial wood energy plantations produce most of the energy for the iron and steel industries. The wood serves not only as an energy source, but also as a reducing agent for the iron; that is, it replaces coke as well as coal. Unfortunately, many of the plantations were established on cleared natural tropical forest areas, thereby disrupting complex ecosystems.

In the developed world too use of wood for energy has been increasing, initially a response to the oil price rises of the 1970s. In the United States, wood now provides 2–3% of energy requirements and is therefore on a par with nuclear and hydro sources (Schreuder and Vlosky, 1986, in Mather, 1990). In Europe it accounts for approximately 2% of energy consumption (Prins, 1987, in Mather, 1990). It is unlikely though that plantations in the developed world will be greatly expanded to produce domestic fuelwood, since here there are so many energy alternatives. Nevertheless, they could be established for commercial energy production. Countries such as Finland and Sweden, in particular, which have almost no indigenous fossil fuel reserves, are very committed to the concept of energy forest (Gustaffson, 1987; Siren, 1980).

3.4 Set Aside

In EC countries the introduction of "set aside" grants is encouraging the establishment of trees on land taken out of agricultural production. To avoid contributing further to agricultural surpluses, these grants do not cover the combination of agricultural and tree crops. It has been estimated that over 15 million ha will have to be taken out of food production in the EC if agricultural surpluses and EC expenditure are to be brought under control (Hummel, 1989).

But the land which is being planted up is often "medium-quality" agricultural land. This tends to be land which has not been subject to intensive agricultural use and which is therefore often of considerable environmental value. Many 'traditional' landscapes and ecologically complex habitats could thus be lost if planted with trees. Moreover, such planting is not forestry driven, but rather a 'buffer' for agriculture. So if demand for agricultural land later increases, these new plantations will probably be replanted with agricultural crops. This does not represent a good approach to land use since it does not involve selecting the best possible land use for an area and planning accordingly.

It is also possible that set aside planting will detract attention and resources from the rehabilitation of degraded European forests, a tendency that has been denounced by farmers unions in countries such as Spain and Portugal. However, the EC is now placing emphasis on management as well as establishment of plantations.

3.5 Poles and other Domestic Products

Analyses of forest product use in developing countries have generally over-emphasized fuelwood and other energy needs and ignored the fact that in a great many cases, farmers, villagers and urban dwellers seek building materials first, especially poles and sticks for wattle and daub type construction. But it is now being realized that by producing the latter materials on plantations, some of the pressures on natural forests are removed. This has been seen throughout the highlands of Africa. In Ethiopia, for example, closely planted *Eucalyptus* spp. such as *E. camaldulensis*, and in the cooler regions, *E. globulus*, have become the principal source of construction materials.

The enormous demand for fencing materials—for containing livestock, and for protecting fields, vegetable gardens and compounds against browsing—has been similarly neglected. As in the case of building materials, plantations can meet at least some requirements. Handicraft materials also depend on woodland products and some species, such as African ebony (*Diospyros mespiliformis*) in Tanzania, which is valued for its heartwood, are under pressure as a result. However, plantations of high-value hardwoods for handicraft production can be extremely difficult to establish (see Section 6.1) and the only solution at present is wise use of the remaining resource.

3.6 Watershed Protection and Management

Plantations are often established to improve watershed management and protection. Tree cover protects soil, reducing break-up of soil aggregates and hence erosion, and improves infiltration which reduces runoff. In Honduras, the Integrated Watershed Management Project, one of many such examples, set up jointly by the Honduran State Corporation for Forestry Development (COHDEFOR), FAO, UNDP and the Honduran-German Food for Work Cooperation, reforested hillsides with fast-growing species. Because a new supply of wood was created, the existing natural forests were exploited less intensively and therefore able to fulfil the function of watershed protection (Camino Velozo, 1987).

But although tree root shear strength maintains slope stability, if there is no litter layer or understorey, erosion of soil particles may be greater inside than outside a plantation. This is because the kinetic energy of throughfall can be higher than that of rainfall, outweighing the effect of interception. Shrubby vegetation is therefore sometimes more appropriate for reforestation of a watershed. It too can fulfil a protection function, providing equal or greater soil cover and is suitable for moderately sloped areas. Furthermore, evaporation is not as great from shrubs as it is from trees since tree canopies are rougher and have a greater surface area than short vegetation and so intercept more rainfall. This can be an important consideration if water availability is a problem, or if water yields are important for other uses. However, trees can draw water from greater depths than shrubs (see also Section 7.5).

3.7 Rehabilitation

Tree cover can be reestablished by protecting a degraded area from cutting, fire and grazing. But this is often a long-term strategy and the need for an economic return from the land may make it infeasible (Adlard, 1992a). Instead, hardy exotic species and/or species with roots which have symbionts that can fix nitrogen, such as *Alnus*, *Casuarina*, *Leucaena* or *Robinia*, can be used to speed up the process of soil development through litterfall and the build-up of organic matter, soil organisms and microflora. (But see also Section 6.1.) In Vietnam, for instance, huge areas deforested as a result of unsustainable agriculture, and chemical attacks during the Vietnam War, are being reforested with both indigenous and exotic species (Lam Cong Dinh, 1992).

Plantations are also used to rehabilitate wasteland such as mine spoil or industrial areas. In Papua New Guinea, *Eucalyptus tereticornis* has been planted on copper tailings (Hartley, 1977, in Evans, 1992), on the Jos Plateau in Nigeria, *E. camaldulensis* has been planted to reclaim tin mine spoil (Alexander, 1989), and in the Dominican Republic, *Pinus caribaea* var. *hondurensis* has been planted on opencast iron and nickel workings (Evans, 1992). Areas reclaimed in this way are relatively small though.

Tree planting to rehabilitate areas denuded of cover can also reduce some of the pressures that are leading to deforestation, since the new cover can provide, among other things, an additional if limited source (due to low yields) of woody material or fodder.

3.8 Soil Protection/Erosion Control/Dune Stabilization

A living root system affords soil considerable protection, while the tree biomass above ground can reduce wind speeds and the energy available to cause wind erosion. The rate of soil drying is also reduced (Evans, 1986a). Major tree planting for soil protection has been undertaken in numerous countries, including Cuba, Ethiopia, Indonesia and the Philippines. In Southern China, afforestation of hillsides has incorporated both soil protection and timber production. Positive knock-on effects of 'protective' planting may include increased agricultural yield. However, some claims have been exaggerated—for example at Guesselbodi, Niger. Plantations have also been established to help prevent desert encroachment, for instance in Algeria and Sudan (Evans, 1992).

Sand dune stabilization and fixation are further common reasons for establishing plantations, as has occurred in China, France, Somalia and Vietnam. But a distinction should be drawn between dunes which are the result of disturbance, mismanagement or pressure on arid and semi-arid environments, and others such as many coastal dunes, which are the result of natural processes and of significant environmental value (see the Habitats Directive of the EC). Following this distinction, afforestation of dune areas should be subject to careful prior evaluation.

It also appears likely that tree planting sometimes addresses symptoms rather than causes. In which case, more attention should be paid to the reasons for environmental degradation (Taylor II and Soumaré, 1984). Some commentators have even suggested that the extent to which tree planting can effectively halt or reverse desertification of semi-arid lands, has not been researched adequately (Good *et al.*, 1992). The potential role of improved management of natural vegetation, such as savanna and scrub, requires further investigation too.

3.9 Production of Non-timber Forest Products

Many plantations are grown for non-timber products such as fodder and food. Fodder plantations are particularly important in areas where they reduce foraging pressure on existing woodlands.

There are numerous examples of non-timber plantations, including *Acacia senegal* plantations in Sudan and Senegal for the production of gum arabic (as well as provision of fuelwood and maintenance of natural ecosystems) (Dione, 1987); tamarugo (*Prosopis tamarugo*) algarrobo blanco (*Prosopis alba*) and algarrobo (*Prosopis chilensis*) plantations in Chile for fodder, (Ormazábal, 1991) and bamboo plantations in central Thailand for bamboo shoot production (Chuntanaparb *et al.*, 1985). A common non-timber use for small plantations in many parts of Central America is as "living fences", often incorporating useful species such as *Gliricidia sepium* and *Leaucaena* and *Erythrina* spp. *Gliricidia sepium* is also often grown to produce light poles, particularly in West Africa for yam stakes.

The neem (*Azadirachta indica*) plantation on the plains of Arafat in Saudi Arabia is rather more unusual than those described above. Established on 10 km^2, it provides shade for the two million pilgrims who camp there each year for the Haj rites. It is hoped that the plantation will also be used to provide pest control materials and ingredients for pharmaceutical products (Ahmed *et al.*, 1989).

3.10 Support of Agriculture

In developing countries in particular, plantation forestry makes significant contributions to agriculture. For example, farm planting can improve soil fertility through nitrogen fixing, litterfall, or the addition of mulch to the soil, and provide fodder for livestock and a source of additional income, while farm shelter belts can lead to higher agricultural yields by providing shelter for crops and/or livestock from wind and heat. (See numerous FAO publications.)

3.11 Carbon Fixation

Carbon fixation is a new use for plantations and cannot yet be fully evaluated, although the rate of carbon uptake of different plantation types could probably be calculated more accurately. This rate varies considerably, depending on the species used, the amount of nutrients, water and light available at a particular site, the planting method employed, the survival rate of the trees, management levels, occurrence of fire, and the level of disease and insect attack (Centeno, 1992). The carbon stored depends on the ecosystem the plantation has displaced and what the wood produced is used for.

It is possible that plantations will be established with the sole intention of locking up carbon, but there are various arguments why this would be undesirable (see Section 10). However, a CARE/WRI project in Guatemala, one of the first forestry sector responses to climate change, also incorporates agroforestry and woodlots, and credits for prevention of logging damage. It is funded partly by Applied Energy Services, a US electricity company, to compensate for the carbon dioxide emissions from a new power plant. It is expected to fix 18.1Tg of carbon over a 40-year period (Trexler, 1990; Trexler *et al.*, 1989.).

Similarly in the Netherlands, the four regional electricity companies, collectively known as SEP (Samenwerkende Electriciteitsproduktiebedrijven/Dutch Electricity Board) have committed themselves to allocating around US$10 million each year to establishing at least 150,000ha of tree plantations, for an initial period of 25 years. The plantations are to offset the CO_2 emissions of a 600MW coal-fired Dutch power plant, but it is also intended that they will contribute to the projected development of the regions in which they are established. SEP has created the FACE foundation (Forests Absorbing Carbondioxyde Emission [sic]) to initiate appropriate forestry projects, both in Europe and the tropics (FACE, 1991; van Ginneken, 1991).

The Australian "One Billion Trees Plan" set up in 1989 is also multipurpose in design, aiming not only to restore degraded forests and conserve biodiversity, but also to offset climate change. It envisages planting 1 billion trees on 1 million ha by 2000, including 400 million in community planting and 600 million in natural regeneration and direct seeding, plus a National Afforestation Programme to establish hardwood plantations. The carbon fixation potential of the programme has been calculated at 300–500 million Tm over the 50-year lifetime of the programme (Eckersley, 1989, and Hawke, 1989, in Andrasko 1990a).

4. HISTORY AND EXTENT OF PLANTATIONS

Tree planting in itself is not a new phenomenon. Indeed, the technical ability to plant trees has existed for at least 2,000 years (Mather, 1993). For example, in parts of the ancient Mediterranean world both timber-producing and crop-producing trees were planted at least as early as 255 BC (Glachen, 1967, in Mather 1993). In the Orient commercial tree planting in China took place prior to the sixth century AD (Richardson, 1990, in Mather, 1993). And in Europe, willows and poplars were cultivated over fairly large areas as from the Middle Ages. (See, for instance, Gabbrielli's discussion (1990) of early Italian forestry activities.) It is rather the scale of planting that is new.

4.1 Temperate Regions

By the early 19th Century most European forestry texts included papers on plantation establishment for fairly large tracts of land. So from the 19th Century onwards, planting was carried out both in traditional forest areas and on bare land. For example, *Pinus pinaster* was planted extensively in Les Landes in France. But this new type of forestry probably developed most rapidly in Germany (Savill and Evans, 1986).

Generally speaking, though, it was not until after the Second World War that plantation cover in Europe and other temperate regions became extensive. Since then significant expansion of tree cover through artificial regeneration has occurred in Bulgaria, Finland, Hungary, Portugal, Spain and Sweden, while in Japan, a programme involving reforestation of approximately 10 million ha reached peak planting rates of 400,000ha a year in the 1950s and early 1960s (Sedjo, 1987; Tsay, 1987). In the United States, the agricultural Conservation Reserve Program, introduced in the mid 1980s, aims to plant up millions of hectares of farmland (Mather, 1990).

Plantations in the Northern hemisphere are usually established on land that has not been under forest cover recently—afforestation, as opposed to reforestation. This is the case, for example, in the southern United States and the United Kingdom. Even so, until the early 1980s, natural and semi-natural cork/holm oak forests in Spain and Portugal were being cleared to make way for *Eucalyptus* and *Pinus* plantations.

Most plantations outside the tropics have been planted for industrial purposes, with 75% of the species being long fibred (conifers) and intended for pulp and paper production.

4.2 Tropics and Sub-tropics

For the tropics and sub-tropics, development of large-scale plantations can be traced back to the 16th and 17th Centuries when European expansion and exploration began to take place. Plants and animals were studied and collected, and specimens shifted around the globe. Plant introductions often failed, although teak was introduced successfully into Sri Lanka as long ago as 1680 (Evans, 1992). Teak plantations were later started in India (1840), Myanmar (1856), Pakistan (1866), Bangladesh (1871), Java (1873), Vietnam (1908) and the Philippines (1910) (Adlard, 1992b). Introductions of numerous other tree species were made during the 18th and 19th Centuries as colonization and resettlement in the tropics proceeded. Prior to 1900, though, there was little need to plant trees extensively as an industrial timber resource (Evans, 1992).

The 1900–1945 period saw the first extensive plantings of tree crops in the tropics and sub-tropics, principally in areas or countries with little natural forest that could be used for timber production, and where there had been an early influx of European settlers (Evans, 1992). For example, industrial plantations were established in Brazil from the early 1900s, particularly in the state of São Paulo, to supply wood for railway ties, and by 1945, South Africa had 180,000ha of plantations of *Pinus patula*, *P. elliottii*, *P. taeda* and some *Eucalyptus* spp., in the Eastern Transvaal (Evans, 1992).

As they matured, species trials and some of the older plantations demonstrated the silvicultural potential for growing tree crops on a large scale in these regions. This factor, combined with the formation of international development agencies interested in industrial forestry as a means of promoting economic development, led to a great expansion in plantation area. Direct private investment in plantations also played a role. Between 1966 and 1977, the yearly planting rate in Brazil alone rose from 40,000 to 500,000ha. Between 1965 and 1980, plantations in the Congo, mainly of *Eucalyptus*, increased ten times to 15,000ha, and about six times in the Sudan to 180,000ha. Between 1966 and 1979, India established around 2.8 million ha of new plantations. A total of nearly 192,000ha of all types of plantation are reported to have been established in the Philippines in 1990. And currently, Cameroon, Colombia, Peru, Vietnam and Zaïre have annual planting programmes of 5,300ha, 170,000ha, 16,000ha, 150,000ha and 11,000ha respectively. Other countries which have seen the development of extensive plantation areas include Australia, Chile, China, Indonesia, Korea and New Zealand (Ball, 1992; Evans, 1992; Mather, 1990; Rao 1991).

But as in temperate regions, many large-scale plantations in the tropics and sub-tropics have been criticized for their negative environmental impacts. Huge areas of natural forest have been cleared to make way for them and mismanagement has occurred, greatly contributing to soil problems such as erosion and nutrient depletion. Moreover, financial benefits have often been less than projected. Consequently, many of the major financing institutions, such as the World Bank, have reconsidered their forestry lending policies and now focus more on small-scale plantations.

Grassland and scrub bushland areas, despite their environmental values, are commonly used for large-scale plantations in the tropics and sub-tropics. Almost all Brazilian plantations, for instance, are concentrated in grassland and scrub bushland areas from which forests have long since disappeared. As Lanly (1982) has noted, the majority of reforestation programmes are not carried out where deforestation is currently taking place. Mather (1990) develops this slightly further, pointing out that most deforestation is concentrated in low latitudes while much reforestation is in the temperate or sub-tropical zones. This perhaps suggests that forestry is not focusing sufficiently on the areas where it is most needed. Care must be taken with this line of argument, though, since natural forests and plantations are not directly comparable. And few would wish to see tree planting as the main means of regenerating natural forest.

4.3 Area

Knowledge amassed concerning the history of plantation establishment is not matched by precise knowledge of extent. Mather (1990) laments that there is no convenient source of data for plantation areas. Moreover, figures relating to plantation extent are not always comparable since different dates and definitions of what constitutes plantations are used. (See FAO, 1967, for a full discussion of definitions.) Compilations of plantation figures are sometimes unreliable too. Reijneveld (1983) has pointed out many of the discrepancies in estimates between various authors.

Gauthier (1991) complains that while statistics on areas planted in the United States and developing countries exist, little is known of either the extent to which planting is successful or unsuccessful, or how often plantations are harvested. Government forestry extension programmes are a case in point. Figures referring to numbers of seedlings distributed and planted are often available, but not figures referring to survival rates, which are frequently low due to lack of follow-up management. Yet without such information it is impossible to gain a true picture of plantation extent (Gauthier 1991; Mather 1990). Figures given for annual plantation establishment can therefore be regarded as approximations only. Ball (1992) comments that data is insufficient not only in terms of monitoring plantation programmes, but also for evaluation of their environmental and social impacts.

Logan (1967) estimated that there were approximately 81 million ha of plantations in 1965. Persson (1974) came up with a figure of 95 million ha for the early 1970s and Evans (1986a) estimated plantations to have attained an area of 120–140 million ha by the mid-1980s. Evans has also estimated that there are around 42 million ha of plantations between 27°N and 27°S of the equator (Evans, 1992). Current estimated total figures are around 100–135 million ha; 75% of plantations are believed to be in temperate regions and 25% in the tropics and sub-tropics. But this can be expected to change, with the proportion of plantations in the tropics and sub-tropics increasing. Current annual plantation establishment and reestablishment are estimated by Gauthier (1991) and Mather (1990) to be approximately 1.2 million ha in the tropics and 10 million ha in the temperate zones. For the tropics, Ball (1992) has come up with the rather higher figure of 3 million ha/yr for the decade 1981–1990. He attributes the increase to more tree planting through community forestry and fuelwood programmes. Tropical deforestation, on the other hand, is currently estimated at 16.9 million ha, or 0.9%, a year (FAO, 1991).

Gauthier (1991), who estimates that plantations cover a little over 100 million ha, or 2.6% of the world's total forest area, has calculated that plantations now provide between 7 and 10% of current world industrial roundwood. As plantations established in the 1960s come on stream, these percentages will rise. For example, in the year 2000, Australia, Chile and New Zealand, could supply around 77 million m^3 annually from their pine plantations as opposed to their current 35 million m^3.

Gauthier (1991) has noted the imbalance in the fact that the tropics contain only 16% of the world's plantations but 75% of the world's population. He further notes that although fuelwood consumption may rise significantly in this region, growth in demand for products made from industrial wood is likely to be slow, since any increase in disposable income will probably be slight. This is a factor which should be considered when proposals for establishing large-scale industrial plantations in the tropics are put forward as a means of offsetting carbon dioxide emissions.

Lanly (1982) reckons that 60% of the world's plantations are industrial and 40% non-industrial. But this can only be regarded as a very rough approximation at best and is not particularly useful given the extremely varied reasons for which and means by which plantations are established. Moreover, this ratio is probably changing given that more tree planting is now being carried out for non-industrial purposes (although the bulk of this is for small plantations).

5. CRITICAL ENVIRONMENTAL ISSUES

The environmental, social and economic impacts of plantations are of considerable concern to the conservation community. Some of them are hard to separate, for instance, those pertaining to soil and water. Others depend as much on the characteristics of the surrounding area as on the nature of the plantation itself. For example, a large-scale industrial timber plantation established on communal areas which previously met the subsistence needs of local rural poor, will have far greater negative social impacts than a similar plantation in Sweden, sited a considerable distance from populated areas. Yet the negative environmental impacts of the latter may be greater if the plantation was established on an area of high conservation value rather than on degraded land. Clearly, plantations must be evaluated individually, and preferably from a broad range of perspectives.

In this report, environmental impacts are considered under the headings of species, soil and water, pests and diseases, and biological diversity. Social and economic impacts are more diffuse but considered under the headings of social and community forestry, land issues and economic issues. Climate change in relation to plantations is discussed under a separate heading, since it has become a complex issue in itself.

6. SPECIES

6.1 Indigenous versus Exotic

Indigenous, local, species are native to a region or country. "Region" is understood here as referring to an area within a country or to an area which contains several countries. Conversely, exotic species are species growing in an area in which they do not naturally occur. *Pinus elliottii*, for instance, is indigenous to South Carolina (US), but an exotic when grown in North Carolina (US), where it does not naturally occur (Zobel *et al.*, 1987). However, if exotics were introduced so long ago that they have since become naturalized (see Section 6.3), they are often no longer regarded as exotics. In the discussion which follows, exotics are taken to refer to species introduced since the 19th Century.

Local species have several advantages for planting. They are adapted to the local environment, provide habitat for its wildlife and are able to regenerate naturally. Moreover, they have generally evolved considerable resistance to local insect pests and diseases. So in many circumstances they will be equal to or better than exotics for the purposes specified. Examples might include protective afforestation or situations where conservation of native flora and fauna is a major consideration (Evans, 1992). Indigenous species are also usually preferred by local people for subsistence, environmental or cultural reasons (Bass, 1992). In Sri Lanka, for instance, exotic pine plantations planted to prevent soil erosion are sometimes burned by villagers because many of them were established on former communal areas and, unlike the natural forest, do not provide familiar non-timber products. However, if the aim is to obtain wood, farmers in developing countries often actually prefer to plant exotic trees.

Growth of planted indigenous species may be slower than for an exotic species, but will not matter enormously if wood yield is not of primary importance. But it will matter if it is necessary to attain soil cover rapidly; in which case, careful attention should be paid to initial site preparation and suppression of competing vegetation. If this is done, native species can show much faster growth than they would under their normal natural conditions. *Araucaria cunninghamii* and *A. hunsteinii*, for instance, have been grown successfully in large plantations in Australia and Papua New Guinea, while species of *Terminalia* and *Triplochiton* have adapted well to plantation work in Africa (Shepherd, 1986). But if planted in large monoculture, indigenous species can prove susceptible to disease and predators. An oft-quoted example is Meliaceae, which survives at low densities in natural forests, but in large monoculture may rapidly fall prey to shoot borers. In particular, *Swietenia* spp. are attacked by *Hypsipyla robusta* in Asia, and *H. grandella* and *H. ferrealis* in the Neotropics. This could, however, be seen as an argument for small-scale plantation work, rather than one for the use of exotics.

Unfortunately, in addition to being slow growing, many of the different genera of tropical hardwoods are also very difficult to establish in production plantations. (Exceptions include teak (*Tectona grandis*), and some Asian dipterocarps.) This is largely because their biology is not well understood and silvicultural knowledge concerning how to handle them inadequate (Zobel *et al.*, 1987). Furthermore, the tropical heavy hardwood species with desirable timber are late secondary or climax species, whereas for large-scale plantations the most useful species have so far tended to be invaders of gaps, pioneers, or early secondary species. These grow quickly and can do so in open areas. Therefore, although plantations of indigenous species do exist in the tropics and sub-tropics, for example of teak and some pines such as *Pinus caribaea*, *P. kesiya*, *P. merkusii*, and *Albizzia falcataria*, they are not as common as those of exotic species. Hardwoods in temperate regions can also prove difficult to regenerate in large-scale plantations, but because of the need for protection against grazing, which is expensive, rather than for silvicultural reasons (Evans, 1985).

The faster growth of exotics—one of the reasons cited most frequently for planting them—is attributable mainly to their inherent genetic make-up, the pioneer-type characteristics of most of them, and the often more fertile soil of the sites where they are planted. A secondary though still important factor is absence of the pests, diseases and competitors found in the species' native environment. In the tropics rotations may be as short as four to seven years for species such as *Eucalyptus* and *Gmelina*. Growth rates, however, are only one aspect of forestry. In some instances a high-quality slow-grown timber (for use in cabinet making) is more desirable than a poor-quality fast-grown timber (suitable only for pulp). In yet others, dry weight (for fuelwood uses) will be the most important characteristic.

Elsewhere, use of exotics is high since the indigenous species available for plantation establishment appear to be relatively few. (But see Section 6.4.) Northern Europe, for instance, has a relatively poor vascular flora, with only five native coniferous tree species north of the Alps. Species such as Douglas fir (*Pseudotsuga menziesii*), Sitka spruce (*Picea sitchensis*) and Loblolly pine (*Pinus taeda*) are therefore commonly used instead for production plantations. Similarly, in discussing the establishment of fuelwood plantations in Ethiopia, Pohjonen and Pukkala (1990) note that suitable productive pioneer tree species have not (yet) been found in the Ethiopian flora. The Ethiopian National Forestry Service accordingly recommends exotic pioneer species such as *Eucalyptus* and wattles (*Acacia decurrens* and *A. mearnsii*) for short-rotation forestry.

Species choice may also be limited by seedling availability. Seedlings of local, native species are often not obtainable in sufficient quantity for establishing a large-scale plantation. There is evidently a need for greater research into methods for multiplying seedlings. (See also Section 6.4.)

Technology is another reason cited to justify the use of exotics. In the tropics and sub-tropics, large-scale exotic conifer plantations are often established, partly in response to the greatly increasing need for wood for industrial use. But it is also because the technology for the growing and processing of some coniferous species has been developed further and is readily available, whereas the reverse is true for many of the indigenous species of these regions. In particular, the high density of many tropical woods and their wood properties, such as high silica content and heterogeneity, create processing difficulties. Further technological research could probably overcome some of them.

Hardy exotics, particularly pioneers, are often favoured too for rehabilitating degraded land. It is commonly believed that indigenous species could not survive in the drastically changed environment that typifies most degraded areas, or improve the soil in the way that an exotic can. In the southern United States, for instance, soil productivity increased following the establishment of exotic pine on piedmont soils that had been eroded by farming (Zobel and Davey, 1977). It is possible, however, that this belief merely reflects lack of research into use of native species. Sometimes too, foresters 'forget' that other vegetation such as grass or shrubs can be more appropriate for rehabilitation purposes.

If they are to be used, exotics should be selected with considerable care and efforts made to ascertain that they will integrate into their proposed new environments. Local trials, with good scientific support, are particularly important (Campinhos Jr, 1991). *Eucalyptus grandis*, for instance, is a widely used plantation exotic, but great attention should be paid to site selection since it is rather site-demanding and requires intensive silviculture to obtain high productivity and profitability (Schönau, 1984). But of course, all species, whether exotic or indigenous, should be carefully matched with site.

To maintain productivity, or in case of the need for adaptation due to climate or socioeconomic change, the sources of exotic material must be studied and protected in their native habitat. Yet regions which are a rich source of exotics also tend to be those with high deforestation rates. Central America, which is exceedingly rich in pine and some other species, many of which are used widely in production plantations in South America, Africa, Asia and Australia, is a good example. Likewise, dry zone species used for non-industrial purposes such as fuelwood, are threatened by multiple pressures, including grazing, exploitation for fuelwood, and fires. This correlates with the comment made in Section 4.2 that forestry activities are not focusing sufficiently on areas where they are needed most.

The problem of loss in countries of origin is not restricted to species, but extends to varieties and provenances (Zobel *et al*., 1987). Efforts such as those being made through a government project in Matagalpa, northern Nicaragua, to conserve natural stands of *Pinus caribaea* of high timber quality in order to maintain genetic diversity, must be carried out on a much wider scale if the productivity of large-scale industrial plantations is to be assured.

6.2 Large-scale Monocultures versus Mixed Species

Large-scale monocultures are often planted because of industrial requirements for considerable supplies of a uniform and dependable product. It is much easier and cheaper to produce this from single species plantations than from mixed plantations or mixed and/or uneven aged forests. Operations such as planting, tending, thinning and clearfelling can be regularized, and the relevant instruction and training of employees carried out more easily. Moreover, great familiarity with a single species usually enables more rapid detection and identification of problems such as disease and nutrient deficiency. A greater number of tree species, on the other hand, means increasing funds assigned to plant protection or giving less attention to some species (Chou, 1981). Besides, if pioneer species are used, they may not be adapted to growing in mixtures.

But monocultures are frequently criticized, and not only on environmental grounds such as increased susceptibility to fire, and pest and disease problems. For example, plantations comprising a single exotic species are often regarded as a particularly alien and unattractive element in the natural landscape. (See also Section 8.4 and Section 12.1.3.)

Bass (1992) therefore suggests that in many parts of the world plantation developers will be increasingly obliged to seek a compromise whereby they supplement their single species/purpose schemes with land set aside for activities such as agriculture. Alternatively, they will have to plant for multiple purposes and therefore with several species. Surprisingly perhaps, the beneficial synergistic effects of mixing species, as demonstrated by agroforestry, appear not to have been considered seriously in connection with large-scale plantations. To date most studies on mixed plantations have been carried out in temperate zones and have not included controlled experiments to provide evidence of improved growth rates, out-turn, financial yield or ecological benefits (Ball, 1992).

However, a study by Wormald (in preparation) with special reference to the tropics and sub-tropics finds that there are certain situations where the forest manager or planner should consider the use of mixtures, including:

- community forests where more than one species in a mixture helps to meet multiple end uses and acts against failure of one of the species;

- firebreaks to create a discontinuity in pure crops, or in intimate mixture, to introduce less flammable fuel;
- nurse crops, especially for the reduction of insect attack;
- encouragement of the shrub and herb layer, through regular thinning and the avoidance of overstocking.

6.3 Problems with Exotics

As many exotics are ecological pioneers or from early-successional stages, their colonizing characteristics have on occasion enabled them to compete successfully against native vegetation for resources. This has been seen in South Africa and New Zealand, where exotic pines such as *Pinus patula*, *P. pinaster*, *P. radiata* and *P. strobus* are suppressing indigenous forest species (Zobel *et al.*, 1987). Similarly, *Azadirachta indica*, introduced from Asia into Africa, has been spread by birds into indigenous woodlands (Davidson, 1987). Such 'invasions' can be a great problem for parks and reserves since they can disrupt the balance of ecosystems and threaten the integrity of the areas in question.

Another problem, albeit rare, mentioned by Quezel *et al.* (1990) is that of "genetic pollution" due to hybridization. This may result when exotic species are introduced into a region which contains native species to which they are closely related. Quezel *et al.* (1990) cite *Pinus nigra* subsp. *laricio* which is mixing with the native populations of *P. nigra* subsp. *salzmanni* in Languedoc and Roussillon, France. Provenance hybridization within the same species has also been known to occur. It is therefore important that the integrity of native provenances is maintained and not polluted through moving species around unnecessarily and careless handling and identification.

One reason for not introducing exotics is that their impact on native fauna and flora cannot be fully evaluated until they have been introduced, and once introduced they may prove impossible to eradicate. Against this, however, it can be stated that in some cases, such as the long-ago introductions of sweet chestnut (*Castanea sativa*) and sycamore (*Acer pseudoplatanus*) into Britain, species have integrated well into their new environment and become naturalized. Only purists would argue that they do not contribute to the landscape. (See also reference to *Nothofagus* in Section 9.1.3.)

6.4 Species Choice

Species choice for plantations is very restricted. Evans (1987) has noted that in the tropics, about 85% of all plantation forestry is dominated by two genera (eucalypts and pines) and one species (teak). The eucalypt and pine genera are large, but the focus tends to be on particular species within them. A few species also predominate in large-scale plantation forestry in the sub-tropics and temperate regions. In Chile and New Zealand production plantations are composed mainly of *Pinus radiata*, while in Britain and Ireland, *P. sitchensis* and *P. contorta* are commonly used at the higher latitudes (Mather, 1990). This situation is exacerbated by the fact that species research for production and management purposes concentrates on less than 1% of tree species and among that 1%, primarily on species for timber production.

Even in agroforestry and social and community forestry, attention has focused on multipurpose, 'wonder trees', tending to distract attention from use of other species. Evans (1992) mentions select, highly productive strains of *Leucaena leucocephala*, introduced everywhere and often on unsuitable sites. Although the species has many useful qualities, it is not universally appropriate.

The same may be said of neem (*Azadirachta indica*) which has been favoured as the 'workhorse' of many plantation schemes in the Sahel (Taylor II and Soumaré, 1984).

Admittedly, large-scale monoculture plantations can be worthwhile financially, since their wood characteristics meet certain requirements. But efforts to diversify research on both native and exotic species would enable a greater range of needs to be met, for example, on social forestry projects, and also contribute to nature conservation. Whitmore (1987) refers to Latin America, but his argument can be extended to all regions, when he suggests that more attention should be paid to using native species in plantations and to their genetic improvement, and especially for sites where exotics have failed. Montagnini and Sancho (1990) referring to the lowland humid tropics similarly state that there is a need for more information on the performance of native tree species grown in plantations or agroforestry systems. Fearnside (in press), in discussing development of agroforestry systems to rehabilitate degraded lands in the Amazon, suggests using indigenous knowledge to help identify suitable species. In this instance, favoured species would be those which produce high unit value products with relatively low nutrient demands.

7. SOIL AND WATER

Large-scale plantations are sometimes accredited with negative impacts on soil and water. This is not so much the case with small-scale plantations, for example those established in the tropics by and for the benefit of local people. There are at least three reasons for this.

Firstly, these smaller plantations have simply not been subject to as much environmental evaluation, perhaps because social issues such as community participation have attracted more attention. (Conversely, assessment of large-scale industrial plantations has been of environmental rather than social impacts, despite the fact that the latter may be as or more significant than the former.) Secondly, they are less likely to disrupt natural processes to the same extent as large-scale plantations, since methods used to establish and manage them are generally not as intensive. Thirdly, even if a small-scale plantation is planted as intensively, its effects will be less far-reaching because of its smaller scale.

For example, preparing the ground for a large-scale plantation is likely to entail using heavy machinery. Methods used to establish small-scale plantations in the tropics, on the other hand, usually rely on local labour using rudimentary tools, which have a much lower impact on the soil. Similarly, harvesting of large-scale plantations tends to be mechanized, whereas harvesting of small-scale plantations is more often carried out with animals such as oxen or mules. This section accordingly refers mainly to large-scale plantations.

However, in certain cases, some of the problems afflicting large-scale plantations can affect small-scale plantations. Thus many small-scale plantations established next to each other, and all growing the same species in order to meet the energy requirements of a local factory, could be subject to nutrient depletion. Large-scale and small-scale plantations are in fact governed by many of the same silvicultural requirements and principles of good management (Ball, 1992).

And although some of the negative soil and water impacts mentioned here occur principally in relation to plantation forestry, for example during site preparation, others are equally applicable to forestry generally, such as those occurring during mechanized harvesting. It also needs to be stressed that it is difficult to reach a definitive conclusion on issues such as nutrient depletion. Not only do plantation sites vary, but many of them have not reached even their second or third rotations.

Finally, it should be remembered that large-scale plantations, are often linked to industrial activities such as pulp and paper production which themselves may have significant negative environmental effects, such as pollution of water sources due to discharge of effluents from mills. Full assessment of many large-scale plantations would therefore entail analysis of the full life cycle of associated forest products.

7.1 Nutrients

Nutrient recycling tends to be more efficient in natural forests than in plantations. Additionally, certain practices can exacerbate nutrient depletion or loss in plantations.

For example, continued cropping, particularly of exotic monocultures, over several rotations, can lead to an overall reduction of site productivity due to an especially high uptake of soil nutrients. The risk is greater if clearfelling or whole tree harvesting, in combination with short rotations, is practised. Whole tree harvesting, which extracts most of the above ground portion of the tree, removes 1.5 to 4 times more nutrients than does bole only harvesting (Kimmins,

1977). Under some particularly intensive systems in temperate zones, the tree roots are also extracted, making rates of nutrient removal and erosion even higher. Short rotations mean that the rotation is little longer than the canopy building stage. In other words, a full nutrient cycle has not been built up prior to harvesting.

Leaching losses will also increase if a site is disturbed often. In fact, nutrient *loss* through soil disturbance may be as great a problem as nutrient *depletion* through extraction of biomass. Leaching can also occur prior to replanting, and prior to canopy closure if the soil has remained exposed due to clearance of vegetation. Energy plantations in particular have considerable potential for loss of site productivity. Practices such as burning can result in nutrient losses too, especially if not well controlled.

One of the disadvantages of some pioneer species is that their litter may take much longer to decay thereby preventing the return of nutrients to the soil. This is especially true of conifer and *Eucalyptus* plantations (Brand *et al.*, 1986; Miller, 1979) in temperate regions. For *Eucalyptus*, knock-on effects of non-decay of litter include failure of natural regeneration or an understorey to develop and increased fire hazard, especially in the Transveraria group. In the tropics and sub-tropics, where leaf litter does not accumulate because of rapid decomposition, due to high temperatures, the problem is less severe (Zobel *et al.*, 1987). However, even in the tropics, if certain *Eucalyptus* are used as an exotic, litter decay may be much slower than it would be in the trees' native environment. This is because the specialized organisms that normally aid decomposition of *Eucalyptus* leaves—which have a highly unusual chemical composition—are absent.

Eucalyptus litter can be poor in nutrients anyway, since some species, as part of their adaptation to a nutrient-poor environment, withdraw up to 75% of the phosphate contained in their leaves, prior to abscission (Attiwill *et al.*, 1978). When *Eucalyptus* are grown in plantations fertilizer inputs may therefore be necessary. Fertilizer inputs are also usually required for marginal soils. However, when referring to degraded grassland available for large-scale planting in the tropics, Evans (1987) stresses that such land is well understood silviculturally, as has been demonstrated by the successful *Eucalyptus* afforestation of Brazil's cerrados and pine afforestation of African savannahs.

In some plantations assessment of growth and productivity seems to show that nutrient deficiency is unlikely to become a problem. Evans (1986b) examining six-year old, third rotation *Pinus patula* in the Usutu forest in Swaziland, found its growth rates comparable to those of the second rotation. He also assessed the productivity of 12-year old, second rotation *P. patula* on exactly the same site as he had the previous crop and found that production was not significantly different. Examination of second rotation *P. elliottii* showed poorer height growth, but the decline was not considered statistically significant. In one part of the plantation he did report significant yield decline owing to probable nutrient (phosphate) deficiency, but over most of it there was no yield decline over three rotations, despite poor rainfall in recent years (Evans, 1988).

Elsewhere, nutrient deficiency has been experienced in plantations. Widespread decline in second rotation growth of *P. radiata* plantations in South Australia was reported in the 1960s. For recent crops, however, the problem has been largely overcome by careful treatment of the site at harvesting, increasing the level of silvicultural inputs (Boardman, 1979) and controlling grass competition which was absent when first rotation crops were planted. Likewise, on the sandy soils of Rennick in neighbouring Victoria, early second rotation growth of *P. radiata* was considerably better than that of the previous crop, due principally to the benefits of retaining litter and logging residues, and avoidance of slash burning (Squire *et al.*, 1985). Indeed, Will

(1984), discussing the subject of monocultures in both temperate and tropical regions, especially in relation to soil deterioration, states that where problems have occurred, poor forest management is usually to blame.

Nevertheless, because of the long time spans in forestry production and the difficulties in estimating nutrient losses, much remains to be learnt about nutrition management. Admittedly, the problem of possible decreases in productivity of forested sites is now recognized ("second rotation problem"), and quantitative estimates of nutrient exports have been made for several temperate forest ecosystems. But for tropical forests, matching data are rare and nutrient cycles less understood (Hase and Foelster, 1983). Moreover, very few data are yet available on yields from successive rotations for either temperate or tropical plantations. Dyck and Beets (1987), discussing New Zealand plantations, therefore argue that to date no clear guidelines have evolved as to what soil types are most susceptible to productivity declines or what practices are acceptable for preventing them.

In short, it is too soon for making final conclusions about plantations and nutrient depletion. But in the meantime, forest managers can at least choose silvicultural practices which minimize ecosystem disturbance so that nutrient loss does not become unacceptable (Whitehead, 1982). To this can be added the necessity of matching species with site and of undertaking careful site monitoring to enable early detection of nutrient problems. And to balance the picture, it should be pointed out that it is also recognized that establishing tree cover by planting can intiate the process of soil development, thereby improving an impoverished site. This is especially but not exclusively the case with nitrogen fixing species.

7.2 Soil Changes

Some specific tree species are believed to affect soil quality negatively. Conifers, especially spruces and pines if grown at close spacing with little or no thinning, are often attributed with causing soil acidification and, in susceptible soils, accelerated podzolization (Savill and Evans, 1986). These processes are doubtless linked to a very acid humus and slow rates of organic matter decomposition and nutrient release. These can then lead to immobilization of nutrients in soil organic matter. In other words, the effect of a type of species on soil will depend on the characteristics of the soil in which it is planted (Ranger *et al.*, 1990).

In Europe and Canada conifer afforestation also appears to have increased water acidity. In Scotland, the pH of streams draining afforested catchments has fallen and fish stocks declined as a result (Harriman and Morrison, 1981). The effect could be due partly to the coniferous afforestation itself. It could also be due to the filtering of dry acid deposits from the atmosphere by the trees—due to their proximity to power stations—which rainfall then washes into drainage systems and which are not subject to soil buffering first. Such impacts are most likely to occur if planting is carried out on soils which are already highly acidic or have a low buffering capacity. Matching species with site should help prevent such problems.

7.3 Mineral Fertilizers and Chemical Pesticides

Mineral fertilizers are used to correct nutrient deficiencies, either prior to tree planting, to aid growth, or to replace nutrients lost through harvesting. This is a well established practice but one which can cause serious environmental problems.

It can be difficult to establish an optimum strategy, both in terms of amount applied and how this amount is applied, particularly if more than one species is being grown, or in an area where soils have not been thorougly analysed. For the tropics, mapping of soil types is not well advanced and an understanding of their differences even less so. Over-fertilization can lead to toxicity and growth problems.

On some plantations insecticides are used in great abundance. In much of the Basque Country, for example, *Pinus radiata* plantations have proved highly susceptible to insect damage and disease. Blanket applications of the insecticide Diflubenzuron, which has a benzene base and inhibits the formation of chitin, are therefore often made (Groome, 1991). Since many large-scale plantations are only in their second and third rotations, the long-term effects of such treatments on soil, water and native flora and fauna, and also on consumed products including meat, honey, fruit and vegetables, are unknown.

7.4 Erosion

A properly managed plantation usually contributes to erosion control. However, erosion problems can result from poor species selection. For instance, some species do not produce much litter or, as in the case of teak (*Tectona grandis*), have large leaves which suppress ground vegetation. In both cases, the soil is relatively exposed and any overland water flow is likely to create erosion. Management practices such as burning (since it alters the physical qualities of the soil), creating contours (although this is actually intended to combat erosion and increase infiltration), downslope ploughing (if carried out with heavy machinery), and discing and clean cultivation of firebreaks (le Roux, 1990), can also create erosion and runoff if not managed carefully.

In large-scale plantations where forest roads have not been adequately thought out, some stands may be subject to the regular passage of heavy machines, leading to increased soil bulk density, reduced porosity, decreased moisture storage capacity and lower infiltration rates. Extracting felled timber, by skidding, or winching up slopes, can exacerbate these processes. In areas of high and/or intense rainfall, or in dry areas subject to storms and where rates of vegetation growth are low, runoff and erosion can become severe, compounding problems of nutrient and water uptake (Shepherd, 1986). Runoff and erosion can also become serious if biomass is harvested frequently for energy, or if a site is not planted up soon after it has been harvested or cleared of vegetation.

The effects of erosion may be felt some distance from the plantation area, as well as locally. They relate principally to siltation of streams and rivers. Fisheries, agriculture and irrigation systems, may all be affected as a result.

7.5 Water Yield/Flow

If forest cover is established on previously non-forested areas, water yield can be expected to decrease due to improved soil drainage, higher rates of foliar interception of water, and loss through evaporation (Good *et al.*, 1992).

For instance, during the growing season in South Africa, flow frequently ceases in areas where plantations have replaced natural, non-forest vegetation such as shrub or bush (Wicht, 1967). Likewise in the southeastern United States, reforestation has reduced streamflow over large areas of land (Trimble and Weirich, 1987). Ten large river basins encompassing 54,000km^2 of land were partially (10–28%) reforested between 1919 and 1967; annual stream discharge was found to decline by between 4 and 21%. In Fiji too, extensive reforestation and afforestation programmes of dry zone grasslands have resulted in reduced water yield in streams (Manner and Drysdale, 1981, in Hamilton and King, 1983).

Generally, catchment water yields lower in accordance with the density of the plantation (Nutter, 1979). Of course, in areas where stormflow peaks and volumes are a problem, reforestation or afforestation might be beneficial, since they generally reduce these. However, Bruijnzeel (1989) believes that the influence of undisturbed tree cover on the magnitude of extreme events is marginal at best.

Water yield may also decline following species change. Flow from a coniferous catchment may be as much as 20% less than that from a hardwood catchment. This is due to the greater annual interception loss and greater aboveground and longer lived biomass of conifers. Swank and Douglas (1974) cite a watershed, previously covered in hardwood forest and now planted with white pine. After 15 years, it showed a reduced water yield for every month. In early spring when the hardwood leaves would only have been emerging, the pine needles were already intercepting water and actively transpiring.

Similarly, work in Madagascar demonstrated that old, exotic *Eucalyptus* forest consumed 100 to 200mm of rainfall equivalent more per year than native, mature forest (Bailley *et al.*, 1974, in Evans, 1992). However, Samraj, Chinnamani and Haldorai (1977 in Evans, 1992) found little difference in run-off between plantations of *Eucalyptus globulus*, *Acacia mearnsii* and native shola forest. Obviously, allowances have to be made for the differences in species and sites where studies are undertaken. Nevertheless, the greatest effects are probably due to the conversion process itself, for example when it involves vegetation clearance, rather than to a change in type of forest (Hamilton and King, 1983).

Eucalyptus especially have been blamed for reducing water yields. In dry conditions the stomata of many other species close to prevent evapotranspiration and water loss. But in the case of *Eucalyptus*, deep root systems locate water and continue uptake. In some countries such as India this has caused problems for local communities. However, the argument should not be that *Eucalyptus* generally are unsuitable for planting outside their native environment, but that more care needs to be taken over species selection. Calder (1986, in Good *et al.*, 1992) has concluded that water loss through canopy interception is in fact less for *Eucalyptus* than other trees of similar height grown at similar planting densities. It also appears that eucalypt transpiration rates, except in areas of shallow ground water, may be similar to those of other non-deciduous tree species. (See Poore and Fries (1985) for a full discussion of the ecological debate surrounding *Eucalyptus* spp.)

7.6 Water Quality

Water quality can be affected negatively by plantation establishment since ground preparation, including mechanical ploughing and ditching, often leads to an initial increase in sediment yield. Measures such as terminating furrows and ditches some distance back from water courses and leaving buffer filter strips of native vegetation along watercourses can reduce the problem.

Intensive management practices, including fertilization, application of insectides, chemical brush control, and burning, can also reduce water quality. For example, usually less than half of applied fertilizers, is used by trees (van Hook *et al.*, 1982). This is especially true of nitrogen fertilizers. The excess may be washed or leached into drainage water and cause nutrient enrichment downstream or change the pH of neighbouring waterways. For instance, fertilizers containing nitrogen in the form of ammonium are highly acidifiying (Good *et al.*, 1992).

8. PESTS AND DISEASES

Much of the discussion in this section refers to large-scale plantations, since this is the focus of most of the literature concerning plantation pests and diseases. Small-scale plantations are less prone to these problems anyway, since the trees within them tend to be more widely spaced and more carefully managed. However, some of the principles applying to large-scale plantations can probably be applied equally to small-scale plantations. For example, an exotic in a small-scale plantation is also susceptible to attack by native pests and diseases, or to attack by a native pest from its native environment. Similarly, a single species planted in many small adjacent plots amounting to a large total area, should be of a broad genetic base in order to increase its resistance.

8.1 Even-aged Stands

In even-aged plantations, and natural forests to a certain extent, disease and insect problems can be related specifically to the establishment phase, the sapling stage and the pole/mature stage. However, in terms of incidence and diversity, they can be expected to be greatest in the nursery, due to plant crowding. Conversely, disease and pest control can be achieved much more easily and cheaply in nurseries than in forest plantations. Moreover, problems may be much reduced once the stock has been transplanted to the field (Gibson and Jones, 1977).

During the establishment phase, pathogens derived from the nursery are sometimes transferred into the field with the planting stock. During the sapling stage, intense competition sets in between trees, and pathogens which attack stressed trees may proliferate, making the trees more vulnerable to insect attack too. Competition for water, for example, can result in low bark moisture content for suppressed individuals, rendering them susceptible to infection by facultative agents. Foliage and bark diseases are relatively common at this stage. During the pole/mature stage, forestry interventions such as brashing, leading to wounds, can facilitate the spread of bark pathogens (Gibson and Jones, 1977; Murray, 1979). Even-aged stands may prove particularly vulnerable to attack by insects which favour trees of a specific age.

8.2 Exotic Tree Species and Pests and Diseases

Exotic species have been regarded as both more resistant and more susceptible to pests and diseases. Growing a species as an exotic may release it from its natural pests, thus improving its health and performance (Zobel *et al.*, 1987). It has been claimed that when planted as exotics, *Eucalyptus* are generally free from the leaf-cutting insects which abound in their natural habitat (Pryor, 1978). Such resistance can be eroded through time, though, with native pests and pathogens adapting to the species in question. In East Africa, for instance, the podocarpus moth (*Gonometa podocarpi*) spread to nine exotic species of the genera *Pinus*, *Cupressus*, *Acacia* and *Eucalyptus* (Austara, 1971).

Hartshorn (1983), in discussing tropical plantation forestry, comments that there is a considerable literature on native insects switching from native hosts to introduced tree species, but that reasons for this switch are not well understood. (This is quite different to the problem of erosion of resistance.) He cites Strong's argument that the number of insect pests of cacao (*Theobroma cacao*) is best described as a function of the area under cultivation rather than time since introduction. Exotics can also become a preferred or favoured food source for native animals and birds (Wilson and Johns, 1982).

Native pests sometimes 'follow' tree species into their new environment, for example, via international transport carriers, or through international trade in or transport of plant materials, and like the trees themselves, are no longer subject to their normal control factors (Simmonds and Greathead, 1977).

In Australia, *Phoracantha semipunctata*, a longhorn beetle, attacks only fallen or highly stressed standing trees, but in areas where it has been 'introduced', it is able to attack trees that are only moderately stressed or of generally low vigour. At high density it may even kill more vigorous trees (Speight and Wainhouse, 1989). In the Iberian Peninsula and North Africa during the early 1980s this insect flourished due to periods of dryness and inflicted considerable damage on exotic *Eucalyptus* plantations. Lack of biological means to control it and the insignificant success obtained with chemical treatments, led to both Algeria and Tunisia abandoning plans for several years, for reforestation with *Eucalyptus* (Quezel *et al.*, 1990).

A similar situation to the one in North Africa occurred during the 1960s in Kenya, which was involved in a project aimed at improving the genetic quality of its pine plantations. Grafted *Pinus caribaea* received from Australia was infested with a pine woolly aphid (*Pineus* sp). But by the time this was realized, air currents had dispersed aphids to nearby pine plantations (Owuor, 1991, in Ciesla, 1992). The insect is now widely distributed in eastern and southern Africa.

Pathogens too have been known to follow exotics. The native range of *Pinus radiata* is California. Here, *Dothistroma pini* does not inflict serious levels of damage. But in areas such as central Brazil, central East Africa, Zimbabwe and New Zealand, it has destroyed large *P. radiata* plantations. In this particular instance, the real problem lay with the foresters responsible for selecting the species. They ignored the advice of specialists who warned that the disease could become severe in pine plantations established in regions with warm and moist summers (Zobel *et al.*, 1987). Nevertheless, the example illustrates the dangers operating.

In colder climates, exotic species have been introduced into environments which contain native species which are closely related to them. Examples include exotic conifers in Europe, North America and northern Asia. The relatively long rotations applied in these regions give the pathogens and pests of the native species more time to adapt to their potential new hosts. In temperate climates the risk of frost damage—which can weaken an exotic and reduce its resistance—is also relatively high. Ideally, of course, local trials will have demonstrated the exotic's ability to withstand any extreme climatic conditions which may arise in its proposed new environment.

8.3 Indigenous Tree Species and Pests and Diseases

Plantations of indigenous timber species can also suffer from severe attack by insects, including indigenous ones. One of the most cited examples is the destruction of iroko (*Chlorophora excelsa*) by the gall bug (*Phytolema lata*) (White and Eastrop, 1964), although resistant strains apparently now exist. For exotics, it may take several decades for pest problems to emerge. But in the tropics establishment of large plantations of native species can be expected to be followed immediately by a simultaneous increase in associated consumer populations. This is probably because, in the tropics, many consumers exist for the same species (Perry and Maghembe, 1989).

Native species also fall prey to attack by exotic insects and diseases. Kenya's self-sufficiency in industrial wood production is already threatened by the invasion of the aphid (*Cinara cuppressi*), which occurs in Europe, India, the Near East and possibly North America. But to make matters worse, the pest threatens to spread from plantations of exotic *Cupressus lusitanica* to *Juniperus procera*, an important indigenous cover species in high elevation watersheds (Ciesla, 1992; Nyaga, 1991). And in Europe, the rapid death and decline of cork oak *Quercus suber*, *Q. ilex* and other oak species, currently occurring across parts of the Mediterranean, have been linked with the highly aggressive exotic root pathogen *Phytopthora cinnamomi*. In 1991 more than 1,050 major decline foci were recorded in the approximately 2.2 million ha of oak forests and plantations in southwest Spain. Comparable oak decline has also been found in adjacent areas of southern Portugal, and in Italy, Morocco and Tunisia. Believed to be originally indigenous to the New Guinea–Celebes region, *P. cinnamomi* has become widely distributed by man and is considered responsible for recent dieback of entire *Eucalyptus* forest communities in western Australia, as well as for the mass dying of native chestnuts across the southeastern United States in the early 1900s (Brasler, 1992).

8.4 Monocultures

Plantation monocultures are usually considered to be more vulnerable to attack from pests and diseases than natural forest. Reference to the r–K continuum goes some way to explaining this. "R" refers to the rate of increase in numbers of a population and "K" to the maximum sustainable number of individuals of that species in a habitat. R-strategists colonize and exploit habitats of an ephemeral nature and have a short generation time. They are not competitive, however, relying on the aforementioned characteristics for their continued survival. Species used in plantations are commonly pioneers and have many of the characteristics of r-selected species. Other r-selected species include pests such as locusts and wheat rusts.

K-selected species display opposite characteristics, including the ability to adapt to restricted habitats where there is strong competition for resources. They typically have a long period between generations and complex defence mechanisms. Climax trees are generally K-selected. Because K-selected species tend to have more defence mechanisms, they may be less susceptible to more damaging attacks than r-selected species. The place of a tree on the r–K continuum might therefore indicate its susceptibility to attack (Savill and Evans, 1986).

There are various examples of pests which have built up because single species plantations have offered improved opportunities for feeding and reproduction, and of pathogens favoured by monoculture. *Hoplocerambyx spinicornis*, for instance, a pest normally found on logs and moribund trees of *Shorea robusta* in India and Pakistan, has become a serious pest of healthy, mature *S. robusta* now that it is frequently established in production plantations (Gibson and Jones, 1977). The principle applies equally to native and exotic pests on native or exotic species.

However, the reputed diversity = stability relationship which is so often used in criticizing monocultures is open to question. It has been argued that diversity does not cause stability but is rather parallel to it (van Emden and Williams, 1974), while some mathematical models demonstrate that complex interactions between different species tend to decrease overall system stability. In fact, natural systems may remain stable in spite of rather than because of their diversity (May, 1973, in Lee, 1992). Complicating the issue is the fact that it is difficult to test experimentally the supposed link between diversity and stability, separating the effect of diversity *per se* from that of climate and the co-evolved relationships between organisms within the community (Speight and Wainhouse, 1989).

It increasingly appears too that a single species plantation will be vulnerable if it is of a restricted genetic base, rather than because it is made up of a single species. Rice is a useful example here. Before the appearance of chemical pesticides, it was presumably the crop's genetic diversity that allowed it to be cultivated as a monoculture for so long. For trees we could refer to a plantation of Loblolly pine (*Pinus taeda*) grown from parents widely selected for adaptability and hence not a 'dangerous' monoculture. A dangerous monoculture would be one established from rooted cuttings or clonal material (Zobel and Talbert, 1984). Thus the failure of plantations of para rubber (*Hevea brasiliensis*) in tropical America before and during World War II, after attack by South American leaf blight, may have been due to the narrow genetic base of the planted host population rather than to the fact that the plantations comprised a monoculture (Chou, 1981).

Natural forests where species diversity is low but stability high and which comprise virtual 'monocultures' do of course exist (Whitehead, 1982). The pine forests of the eastern United States, the beech (*Nothofagus*) forests in New Zealand's South Island and in parts of Chile and Argentina (Savill and Evans, 1986), and *Anisoptera polyandra* forests in the lowlands of Papua New Guinea (Evans, 1992), are obvious examples.

There are at least five points that can be made about these oligarchic forests, though. Firstly, they may consist of different age trees, which exploit different soil levels. This provides some protection, since there are insect species which attack only trees of a particular age. Secondly, they may have arisen due to forest fires which occur on a regular cycle and prevent the development of a mixed forest. For example, some conifers regenerate extensively after periodic fires and form even-aged stands. Thirdly, they may have arisen because the soils on which they grow are very restricted and only one species is adapted to them. Fourthly, they are genetically diverse (with a few well-known exceptions such as *Pinus resinosa*). Fifthly, they may have been evolving for a considerable time and provide habitat for rich animal communities, even though low in tree species diversity.

It's worth remembering too that serious diseases also occur in mixed and exploitable natural forest and that some pathogens can attack several host species. *Phytopthora cinnamomi*, for example, has a host range which covers 444 species belonging to 131 genera of 48 families (Newhook and Podger, 1972). Additionally, there are many insects whose origins as forest pests have been wrongly attributed to monoculture. For numerous insect species, either natural or artificial forest stands are suitable for the build-up of populations, and some were already major pests in their native habitat before dispersal elsewhere. Moreover, infestations can be as severe for individual, isolated trees as for plantations (Gibson and Jones, 1977).

But as a final point, it is worth stressing that monocultures do favour the spread of pests and diseases, although it is not necessarily the case that they will occur.

8.5 Clonal Forestry

In the last 10 years considerable progress has been made in vegetative propagation of tree species, both in temperate regions (for example, *Pinus radiata* in New Zealand and *Picea sitchensis* in the United Kingdom) and in the tropics (for example, *Eucalyptus* spp. in Brazil and *Gmelina arborea* in Malaysia). Genetic improvement is gained through clone selection and replication of crops with specific desired characteristics. Yield improvements of up to 50% have been attained.

But clonal plantations—which usually consist of 10–30 different families—may have a narrower genetic base than crops of seedling origin. Potentially, this makes them more vulnerable to pests and diseases. Indeed, the seriousness of the threat has been seen with the high yielding varieties of *Leucaena leucocephala* whose narrow genetic base greatly contributed to rapid and devastating attack by the psyllid *Heteropsylla cubana* throughout much of Asia and the Pacific.

However, all major breeding and clonal propagation strategies recognize this increased risk and seek to minimize it by maintaining established crops of many clones (of 10–30 families, as noted above) in mixture or mosaic, by pursuing a policy of continuous improvement each year whereby new clones are added and old ones discarded, and by using sexual reproduction for the production of new clonal material and as a back-up.

8.6 Silviculture

Pest and disease problems may be due to factors other than use of monoculture or exotic species. These include: an inappropriate site choice, leading to stress-induced changes in the trees, increasing their vulnerability to pests and diseases; use of a poorly adapted seed source; poor silviculture, such as careless pruning and thinning which can leave scarred live tissue open to infection; and inadequate attention to nutrient and water requirements (Burdon, 1982).

For example, in East Africa, the longhorn beetle (*Oemedia gahani*) spread from indigenous hardwoods onto *Cupressus lusitanica* as a result of poor site-clearing practices and lack of control of pruning operations (Gibson and Jones, 1977). Similarly, root decay caused by wood-destroying basidiomycetes such as *Armillaria mellea* and *Heterobasidion annosum* (*Fomes annosum*), sometimes mistakenly referred to as an example of a disease originating in plantation monocultures, is traceable to stumps and root residues in the soil. Inattention to site history and/or failure to undertake precautionary measures, such as selection of resistant species and stump treatment, rather than its later use for a monoculture, is the crucial factor. (From quite a different viewpoint, *A. mellea* is an appreciated mushroom and could be managed to create a resource of high financial value.)

Fortunately, foresters have great flexibility when choosing species for monocultures, which allows a good margin for avoiding serious pest and disease problems (Gibson and Jones, 1977). For example, if an exotic species has been planted and proves susceptible to a locally prevalent pest or disease, it may be possible to switch to a native species. Regular and careful site inspection by forest staff trained to look for signs of damage can also do much to prevent pest and disease problems or their development to catastrophic levels.

But there is little cause for complacency. In the long-term, pests and diseases may build up to far greater levels than occur at present. In Brazil, there are now numerous pests in *Eucalyptus* plantations, and more are being discovered, whereas almost none were reported early this century (Davidson, 1987). Similarly in many second and third rotation pine plantations in southern Africa, there is evidence of increased numbers of potential disease and pest species (Evans, 1986a). Ball (1992) accordingly calls for improved monitoring of insect populations in plantations and greater exchange of information to support quarantine measures.

8.7 Integrated Pest Management

Integrated pest management (IPM) uses a number of alternative or complementary control procedures in a coordinated manner to prevent damage by insect pests. The theory and practice of IPM were initially developed as an alternative to unilateral application of insecticides to solve pest problems in agriculture (Speight and Wainhouse, 1989). Reasons for using IPM in forestry include the high costs of applying insecticides to forests, the difficulties in controlling levels of forest pests with insecticides, and risks associated with chemical control. Speight and Wainhouse (1989) have summarized the procedures available for preventing pest problems through IPM, as given in Table 5. Many of them simply comprise good and timely forest management and as such should be implemented anyway.

IPM is also appropriate when an exotic pest has already established itself and eradication is no longer a viable option. In this instance, development of an IPM system would include monitoring insect populations and damage levels, and accelerated research and action programmes. The latter would aim to: establish host/pest interactions; identify site and stand conditions which favour the insect; evaluate effects of local natural enemies, and estimate potential resource damage. Additional activities might include screening the natural enemies of the insect for possible introduction; searching for genetically resistant or tolerant strains of host trees; evaluating silvicultural measures to maintain host vigour and resistance to attack, and defining areas requiring treatment. IPM has been used to control recently introduced forest insects such as *Heteropsylla cubana* in Asia and the Pacific, *Dendroctonus micans* in the United Kingdom, *Cinara cupressi* and other conifer aphids in eastern and southern Africa, and *Sirex noctilio* in South America (Ciesla, 1992).

Table 5. IPM procedures for preventing pest problems

Methods for Preventing or Minimizing Pest Problems	
Suggested order of priority	
1	Choose appropriate site, tree species or provenance, origin (exotic or native), planting regime (mixtures or monocultures)
2	Use good silvicultural practices, eg sanitation, hygiene, to maintain tree vigour
	Establish quarantine measures against exotic pests; define internal quarantines for those already established
3	Survey for actual or potential pests, assess size and impact of problem
	Routinely monitor serious or potentially serious pest problems
4	Make control decisions based on all available information related to economic, silvicultural and entomological factors
Methods to Control Pest Outbreaks	
5	Silvicultural or ecological techniques, eg selective felling, hygiene, trap trees
6	Biological control with pathogens for lepidopteran and sawfly defoliators
	Biological control with parasitoids or predators where appropriate
7	Insecticidal control using selective chemicals applied efficiently

Source: Adapted from Speight and Wainhouse (1989)

9. BIOLOGICAL DIVERSITY

The direct impacts of large-scale plantations on biological diversity have been more widely studied and are therefore better understood than those of small-scale plantations. Even so, it is reasonably certain that the impacts of large-scale plantations are more severe since they generally cause greater local disruption and may also have regional consequences.

In this section, large-scale plantation is taken to refer to a plantation which has greater similarity with an agricultural crop, than with a natural forest. This should be noted for examples given below when large-scale plantations are compared with natural forests.

9.1. Large-scale Plantations

Biodiversity, whether it refers to species, the genes they contain, or the ecosystems to which they belong, or to all three, is often in direct conflict with a plantation if production of goods (as opposed to services) is the main aim. For example, maximizing timber yields usually involves reduction of species diversity through elimination of pests, predators and competitors (Lee, 1992).

9.1.1 Ecosystems and Habitats

Large-scale plantations are frequently criticized on account of their lack of biodiversity. Yet if a plantation is established on an impoverished site, the new vegetation and associated return of wildlife can increase it. Moreover, large-scale plantations are not equally poor in species or habitat. A plantation may comprise mixed species or a monoculture, or a mosaic of single species blocks of different species. Species used may be native or exotic, the plantation long-established or very recent, and management techniques more, or less, intensive.

In production plantations in New Zealand, for instance, older exotic stands provide a better habitat for native birds than very young stands. This is because the conditions of older stands become more open as they mature, leading to an increase in other vegetation and other habitats (Bull, 1981). In fact, production plantations usually consist of several age classes, so that although there is little diversity within individual stands, the 'forest' as a whole will contain different habitat types, such as open ground, areas of young trees, closed thickets and mature open stands. Additionally, unplanted areas, including roads and tracks, gullies, rocky areas, and firebreaks, and which normally account for one-fifth of the total area, remain a source of habitat diversity (Evans, 1992). (But see Porter, 1990, below.) This diversity may be a poor substitute for a site's original diversity though.

For instance, replacing an area of diverse tropical rain forest in the Brazilian Amazon with a plantation monoculture would lead to a considerable reduction in ecosystem and species diversity, and perhaps even loss, if the relevant species were endemics. Large plantations have been established on tropical forest sites, for example, the Jarí plantation in Brazil. Postel and Heise (1988) have calculated that at least 15% of all plantations in tropical countries were established at the cost of natural closed forests. Recognizing the problems this has caused, some plantation operators, such as Shell International, have made it their policy not to clear natural forests to release land for plantation forestry (Shell, 1990).

But species may also be lost if a plantation replaces not a different forest type (ie natural forest), but a different habitat altogether. The breeding sites of the Blue Swallow (*Hirundo atrocaerulea*), South Africa's most endangered bird species—which occurs and breeds in the Natal midlands and eastern Transvaal highveld—frequently coincide with prime afforestation sites (Cooper, 1990). Other South African wildlife species, also dependent on natural grassland, are similarly threatened. When a site is afforested, different plant and animal life will colonize the 'new' habitats. But they may not be those originally characteristic of the area or what Good *et al.* (1992) term "natural biodiversity".

Reforesting or afforesting surrounding or adjacent areas also affects species and habitats. Porter (1990), discussing afforestation in Natal, South Africa, points out that sensitive botanical areas include rocky areas of shallow soils, often on steep slopes and ecotonal or transition areas between indigenous forest and grassland, and between grassland and riparian vegetation. Yet these are sometimes included as firebreaks and undergo detrimental annual burning. Porter mentions in particular certain endemic plants associated with sandstone areas that have been incorporated into firebreaks.

Conversely, areas unsuitable for planting trees are often left with their existing native trees, but these may then suffer from an impaired reproductive strategy because fire is excluded from them. Porter cites the many isolated communities of *Protea roupelliae*. If they act as barriers to natural vegetation, plantations can also reduce the dispersal rates of native fauna and flora, thereby increasing fragmentation or island effects. And even if species have the capacity to adapt to the exotic species in plantations, the evolutionary processes will take considerable time. As Barnett (1992) notes, in biological terms the [large-scale] plantation habitat is very recent.

For large-scale plantations, negative effects may extend far and wide. For example, if a plantation disrupts water production from catchments, wetlands and swamps elsewhere may be affected. Afforestation at the headwaters of the Mkuzi, Mzineni, Hluhluwe and Nyalazi rivers, which has resulted in reduced water availability, is one of the principal threats to the St Lucia Wetland Park in Natal. In South Africa generally, it appears that drying up of wetlands due to large-scale afforestation is responsible for the loss of several wattled crane (*Grus carunculata*) nesting sites. The wattled crane is also one of South Africa's most endangered birds (Porter, 1990).

Table 6. The number and percentage of species potentially affected by future afforestation in Natal

	No. of Species	No. of Species in Afforestation Area of Natal	% of Species	No. of Threatened Species*
Fish	62	22	35.5	2
Amphibians	77	57	71.4	1
Reptiles	152	97	63.8	6
Birds**	499	236	47.3	24
Mammals	155	55	35.5	10

* Listed in Red Data Books
** Excludes marine birds

Source: Porter (1990)

9.1.2 Plantation Structure and Plant Diversity

Plantations are less diverse than natural forests because generally they have few tree species per unit and lack structural complexity within stands. (Even a single species fire-climax forest will have greater genetic diversity.) Natural rain forest, for example, has several layers—emergents, upper canopy trees, lower canopy trees, tall shrubs, bushes and ground vegetation, and lianas and epiphytes. But a plantation stand typically consists of one layer only (the canopy). This is because of its even age and because other vegetation will have had little opportunity to grow or been eliminated in order to enhance crop growth. In pine and teak stands, for instance, stocking rates are usually very dense so that little light reaches the forest floor.

Additionally, short rotations or thinning may not allow native herbs and shrubs sufficient time to develop. This is the case in the British uplands where most spruce plantations are harvested at around fifty years of age (Hill, 1979). Native species may also fail to develop because the species used are exotics. In northern Spain, rare native plant species have been shown to be 7.4 times more likely to occur in natural oak/chestnut forest than in a *Eucalyptus* plantation (Bernáldez, 1989). Similarly in the tropics, epiphytes, which are a source of much diversity, take some time to establish themselves and are therefore generally absent from plantations in this region. Lack of old-growth vegetation tends to mean less animal and insect diversity since fewer habitat types are available for colonization. Even within the soil, an even-aged plantation crop will display little diversity, since tree roots will have reached the same rooting depth.

9.1.3 Bird, Mammal and Insect Diversity

Owing to the structural simplicity and relatively few habitat types of large-scale plantations, many bird and mammal species will not use them. Species that do are likely to be common generalists found elsewhere. Bird diversity in particular is linked to vegetational complexity. Homogeneity of canopy, lack of a full understorey or of features such as dead wood, holes and snags, translate into reduced opportunities for feeding, reproduction and concealment. Food too may be less diverse, though not necessarily less abundant.

Numerous examples can be cited of the lower bird diversity of plantations in comparison to natural forests. In the Kinale area of the highlands north of Kijabe in Kenya, indigenous forests of *Brachylaena croton* were cleared to make way for plantations of exotic conifers with significant negative impacts on the native bird fauna in terms of species and densities. (However, the new pine plantations appear to have extended the wintering range for some Palearctic migrant bird species.) (Carlson, 1986). In Portugal, few birds are found within exotic *Eucalyptus* stands since they contain hardly any native insects. This is not surprising given that in native *Eucalyptus* stands in Australia, a fauna has evolved which is specifically adapted to the chemical composition of the species' leaves (Kardell *et al.*, 1986). And in Orissa, India, in an area where tropical deciduous forest had been largely converted to sal (*Shorea robusta*) plantations, only 16 bird species have been observed, even though sal is native to India. For the remaining patches of closed natural forest, the count is 55 species (Ripley, 1979, in Good *et al.*, 1992).

Native mammal diversity is also often reduced by plantations. In Victoria, Australia, some mammalian species dependent on native *Eucalyptus* forests have been able to persist in areas converted to *Pinus radiata*, but only because native forest still exists nearby, thereby illustrating the importance of refuges and corridors of natural vegetation (Friend, 1982). Elsewhere, however, plantations which are protected against intrusion may perform an important conservation function by providing undisturbed habitat for 'sensitive', rare species or species hunted by local people.

Whereas birds and mammals are dependent on both forest structure and tree species, invertebrates tend to be dependent on particular tree species only (Good *et al.*, 1992). In New Zealand, it appears that a mere 10% of the Coleoptera of native forests is found in exotic forests (Bull, 1981). In Sabah, initial results from nine months of sampling showed that for the moths (Lepidoptera) in the super-family Geometroidea, the total number of species captured in monoculture plantations of *Acacia mangium*, *Eucalyptus deglupta*, *Gmelina arborea*, *Paraserianthes falcata* and *Pinus caribaea* var. *hondurensis*, was lower than that of secondary natural dipterocarp forest (Khen *et al.*, 1992). Species richness was still relatively high, though, particularly in the *E. deglupta*.

Occasionally, however, exotic species appear to have greater wildlife as opposed to strict conservation value than some native species. In Britain, *Nothofagus* spp. introduced from Chile as recently as 1910, are exceeded only by oak in the numbers and variety of insects now associated with them. In such cases, though, an important consideration is whether or not the introduced species is well integrated into its new environment.

9.1.4 Genetic Diversity

Plantations can reduce genetic diversity. For example, planting up an area could result in the loss of the last remaining population of a species with a particular genetic make-up. Against this, though, a successful planting programme which seeks to ensure sustainability of production, will include conservation of genetic resources of a broad variation within its strategy. In fact, use of a species for plantation work—which often includes extensive planting and intensive management—is an effective means of conserving its genepool. (See also Section 6.4). In Scandinavia, in the 1940s and 1950s, species of the birch genus (*Betula* spp.) were in demand for fuelwood use. But in the 1960s, birch practically disappeared from many areas due to the timber sector's lack of interest in its properties, and increased consumption of fossil fuels. Then in the 1970s, with the development of new industrial processes, birch was once again in demand. This led to species and provenance testing schemes, selection, breeding and widespread planting.

9.1.5 Management Operations

Plantation management operations sometimes disrupt wildlife. Weeding, thinning and felling disturb habitat, while use of fire can destroy food sources such as leaves, grasses, flowers, seeds and nectar, the specific habitats in which species live and breed, and protective cover. In South Africa the timber industry is prohibited from burning plantation slash between July and October because of the risk of fires getting out of control. Veld burning prior to planting is therefore carried out during the wetter spring and summer seasons, but this means during the growing season for plants and the breeding season for animals (Porter, 1990).

Additionally, fertilizer and insecticide applications affect aquatic flora and fauna, within and outside plantations when waterways become polluted with the minerals or chemicals used. Other operations, involving use of repellents or physical barriers, or more indirect control methods such as habitat manipulation, are often undertaken specifically to reduce certain wildlife populations.

However, some management methods have brought unintended benefits to particular species. Wesley *et al.* (1981, in Good *et al.*, 1992) investigated the effects of excluding deer and of thinning on understorey vegetation biomass and wildlife in plantations in Arkansas and Mississippi, in the United States. Both practices enhanced plant biomass considerably which favoured greater diversity and abundance of small mammals in particular.

9.1.6 Modifying Management

Growing awareness of the effects of large-scale production plantations on biodiversity is leading to the development of management methods which take greater account of wildlife needs. But as with any other plantation objective, that of improving or maintaining conditions for wildlife requires careful definition. For example, since their requirements will not be the same, are some species to be favoured above others? In North America, maintaining habitat for the benefit of the spotted owl (*Strix occidentalis*) means retaining old-growth forests to the detriment of earlier successional stages, although the latter have greater total diversity. In such situations, the answer might be to ensure that other successional stages are allowed to develop elsewhere.

Methods for enhancing the wildlife 'opportunities' of large-scale plantations include:

- Extending rotation periods on specified areas to maintain old-growth timber and encouraging growth of vegetation natural to the area.
- Retaining corridors of the original natural vegetation throughout and alongside plantations, and linked to reserves.
- Leaving ground next to watercourses unplanted (Friend, 1980). Areas which are of marginal production value, such as gullies, stream buffers and steep slopes, are often of high conservation value or 'usefulness'. In Pointe-Noire in the Congo, for instance, riverine forest has been retained in *Eucalyptus* plantations to provide habitat for chimpanzees.
- Allowing an understorey to develop, so that cover and a continuous source of food for animal species are provided, as has been the case for deer in pine plantations in Swaziland, tigers in teak plantations in India, and kangaroos under radiata pine in Australia.
- Managing harvesting and tending operations in such a way that habitat disturbance is minimal.

However, for many species it is not yet known what constitutes the minimum area necessary to maintain a viable population or the type of habitat—for example, open area, fragmented areas or edge—required for long-term conservation. In which case, decisions pertaining to size of corridors and reserves should err on the side of caution (Friend, 1980). The biological ability of forest-dwelling species to respond to intensive management instances is another subject requiring research.

Modifiying plantation management to minimize negative impacts on wildlife usually results in longer initial planning since adequate biological surveys of areas must be carried out, and sometimes also in higher management costs. So it is unlikely that plantation owners, unless legally obliged to do so, would make the necessary modifications of their own accord. In some instances, government financial support could provide the necessary incentive.

9.2 Small-scale Plantations

Little has been written about small-scale plantations and biodiversity. Indeed it is an area which should be investigated further, particularly in view of the increasing emphasis on establishing small-scale plantations in tropical regions.

Lugo (1992), who warns that we should avoid generalizing about all tropical tree plantations or all natural tropical forests, compared small tropical tree plantations with secondary forests of similar age. He found that after 50 years, plant species richness in the understorey of a mahogany (*Swietenia macrophylla*) plantation approached that of its paired secondary forest. It is worth mentioning, though, that Lugo carried out his research in Puerto Rico where the diversity of both primary and secondary forests is reduced on account of Puerto Rico's being an island. Furthermore, the plantation in question had been left untouched for a long period.

Senanayake (1987) writing of home gardens states that those with 40 or more species grown for a mixture of products, including fruit, foliage and fuelwood, may contain avian and soil faunas which are almost as diverse as those of natural forest. In home gardens and agroforestry schemes generally, structural diversity is created by the different vegetation layers and provides many niches for wildlife.

Edge appears to increase biodiversity. (Edge effect is the tendency towards greater variety and density of plant and animal populations in an ecotone.) For example, in Sabah, Malaysia, wildlife was found to be more diverse and abundant on the edge between a 6-year old *Eucalyptus deglupta*/*Gmelina arborea* plantation and secondary forest than in any of the other habitats, primary forest included (Duff *et al.*, 1986). This was probably because the plantation provided additional habitat types not found in the natural forest.

For the same total area several small-scale plantations will have a higher proportion of edge to area than one large-scale plantation and it could be concluded that in terms of numbers, small-scale plantations therefore have a higher diversity value. However, the extra diversity created by edge effect is not the diversity that under pre-plantation conditions would be found in that area. Moreover, in an area where a species or species require large undisturbed or open areas it might be more advisable to have one plantation block, as opposed to several, in order to reduce fragmentation and dispersal effects. There may also be problems of exotic species invading the natural forest where the two are close together, and destruction of the ecotone.

10. CLIMATE CHANGE AND PLANTATIONS

There are various reasons why large-scale plantations have been suggested as means of sequestering carbon. These include the fact that natural forests contain 20 to 100 times more carbon per unit area than agricultural lands (Houghton, 1990a), that many silvicultural techniques focus on maximizing wood yield, thereby fixing maximum amounts of carbon, and the seeming availability of enormous areas of degraded or non-intensively used land, principally in the tropics and sub-tropics, where such plantations could be established. The tropics and sub-tropics are also favoured since tree growth rates there are higher and establishment costs sometimes lower than in temperate regions. In other words, they have "comparative advantages".

10.1 Points of Clarification

But several points (at least) should be made. Firstly, as Houghton (1990a) himself states, the major contributor to the rise in atmospheric carbon dioxide (CO_2) is fossil fuel combustion. Carbon emissions from fossil fuels are currently two to three times greater than those arising from deforestation (Houghton, 1990a). The role that energy conservation could play in reducing climate change would therefore appear to be greater and more appropriate than that offered by forestry alone. It is also one which could be best pursued by the developed countries, who after all are responsible for the majority of carbon emissions and already possess the technology to reduce them. They could also contribute to reducing the CO_2 emissions of developing countries by promoting technology transfer. This would certainly be more equitable, since those responsible for causing the problem would be 'paying' for its solution, rather than vice versa.

Technology transfer could also enable developing countries to reduce their external debts if it led to their reduced consumption of fossil fuels. Conversely, establishing plantations in developing countries to soak up the carbon emissions of developed countries, could become yet another example of exploitation (cf. the dumping of toxic waste) by the North of the South. However, climate change is such an immense problem that emissions reduction technology will not by itself be sufficient to solve it. Additional measures will be required.

Secondly, if current emission trends continue, any forest intervention will be of limited impact (DHV Consultants, 1991). Non-forest strategies must therefore be considered in addition to or instead of forest strategies. Thirdly, forests withdraw carbon from the atmosphere only for as long as they are gaining mass. Having reached maturity, they are approximately in balance with respect to carbon, accumulating it at the same rate as it is released. Forestry options may therefore offer a temporary solution only, which would perhaps bridge 30–100 years (Andrasko, 1990a). Reforestation could however reduce CO_2 emissions indefinitely if wood fuels were to replace fossil fuels (Houghton, 1990a).

Fourthly, the carbon fixation potential of the proposed large-scale plantations would ultimately depend on the use to which the products of these plantations were put. If, for example, the timber harvested was converted to a product which does not have a long-term use, such as paper for printing newspapers, little carbon would be locked up after the growing phase. Once the paper was destroyed the carbon it contained would be returned to the atmosphere. Carbon would be locked up for considerably longer if, however, it was used for furniture or construction.

Fifthly, calculations of the benefits of using plantations to alleviate climate change should be made using a figure for their average biomass rather than for biomass existing at the time of harvest. In 1991 President Collor de Mello announced a proposal to establish 12×10^6 ha of plantations in Brazil. But the calculations of the extent to which they would offset carbon emissions was out by a factor of two, because an average of the biomass was not used (Fearnside, in press).

Sixthly, forests are not the only active pools of the carbon cycle. The oceans actually store far more carbon.

Table 7. Active pools of the carbon cycle

Storage	Amount (Gt)	Percentage
Oceans	36,000	78
Fossil	7,500*	15
Soils	1,500	3
Atmosphere	700	2
Vegetation	560	2
Total	46,000	100

*Estimates range from 5,000 to 10,000

Source: Houghton and Woodwell (1989)

10.2 'Degraded' Area Estimates

Estimates of the degraded land available for forestry vary considerably. Houghton (1990a) has calculated that in Latin America, tropical Asia, and Africa there exist 500 million ha of previously forested lands which are not intensively used and which could be used for forestry. To this figure he adds 365 million ha in the fallow cycle of shifting cultivation, giving a total of 865 million ha. For the purposes of his calculation he assumes that shifting cultivation could be replaced with low-input, permanent agriculture which would allow about 85% of the latter area to return to full-grown forests. Grainger (1990) has estimated the area of degraded land in all three tropical zones with potential for plantation establishment to be only 621 million ha although he considers this a conservative estimate (and has previously quoted a figure of 758 million ha) (Grainger, 1988). Unlike Houghton, he detracts forest fallows on the grounds that they could form part of a sustainable shifting cultivation system. This reduces his figure of land suitable for afforestation to 420–450 million ha. Wiersum and Ketner (1989) base their estimate of 200 million ha on the necessity of reforestation for development purposes.

But the term degraded needs to be treated with caution since it is not well-defined. For instance, it is used to refer to both forest fallow which is degraded relative to mature or primary forest, and weedy grassland. Desertification is also a measure of degradation, but as Grainger (1991) states, there is still no agreement on how to define desertification in terms of measurable parameters. So if realistic programmes are to be drawn up for establishing plantations on degraded lands, much more precise physical measures of soil and vegetation degradation will be required. Without them, other calculations, such as afforestation and reforestation costs, will be imprecise. For example, degraded land undergoing some form of productive use would be less expensive to reforest, and grow more wood, than land which is so degraded that it has been abandoned altogether.

Moreover, what constitutes degraded land to a forester may be regarded as an important habitat by a conservationist, or in the eyes of a subsistence farmer, suitable for crop growing or grazing livestock. In fact, much land reputedly available for reforestation is currently being used by subsistence farmers or nomadic pastoralists to help meet their basic needs. Alienating it for plantation establishment would almost certainly cause social conflict and hardship.

Finally, we may question whether it is worthwhile to establish large-scale plantations on marginal land. Risk of failure is increased and long rotations make it financially unattractive to invest in major land improvements such as irrigation. Interestingly, at a symposium in 1983, foresters were warned that in the next decade they might have to resist pressures for large-scale planting programmes that would be unlikely to be successful for these very reasons (Budowski, 1984).

10.3 Agricultural Land

Calculations of land available for reforestation are based on areas believed to be currently available. That is, they do not take into account the future needs of subsistence farmers for agricultural lands. On the basis of projected population increases, these can be expected to be high. Demand for agricultural land can also be expected to grow since falling commodity prices are forcing many subsistence farmers in developing countries to produce more and more in order to retain the same level of income. Demand for agricultural land will stabilize only if agricultural practices are intensified sustainably, returns for farmers increased and population growth reduced. Admittedly, the rate of population increase is falling in some developing countries, and significantly in, for example, China, Sri Lanka and Thailand. But general intensification of agriculture seems far away given that one of its preconditions is more equitable land distribution.

10.4 Land Tenure, Use and Sovereignty

In many areas of the tropics and sub-tropics, land tenure issues and existing patterns of land use would mitigate against extensive establishment of large plantations. As Sargent and Lowcock (1991) point out, developing countries such as India and Thailand have set their own targets for forest areas and the key problem has been not the setting of the targets, but effective implementation of programmes to meet them. And in 1990, a workshop held in Guatemala to examine the potential for reforestation and restoration in Central America concluded that although suitable land is available, socioeconomic and institutional capacity problems already limit reforestation efforts in the region (World Resources Institute, 1991, in Andrasko, Heaton and Winnett, 1991).

There are also political constraints to large-scale afforestation. Some governments would probably object to participating in a global plantation programme on the grounds that it would contravene regional sovereignty (Grainger, 1991). Furthermore, if large-scale plantations were promoted as part of a global plan to combat climate change, prevailing government policies would need to be taken into account insofar as they encourage or discourage forestry.

10.5 Current and Future Planting Rates

Planting rates in the tropics do not augur well for future massive planting schemes in this region. The average annual *tropical* afforestation rate for 1976–80 was around 1 million ha (Lanly, 1982) and around 2–3 million ha between 1980 and 1990 (Ball, 1992; Evans, 1992). Scrutiny by DHV Consultants (1991) of the National Forestry Action Plans of 15 countries participating in the Tropical Forestry Action Programme, revealed that their future planting will amount to only 2.3 million ha/year. The gap between these rates and the planting of 12 million ha a year recommended by the 1989 Noordwijk Ministerial Conference on Pollution and Climate Change, to "pursue a global balance between deforestation on the one hand and sound forest management on the other", is enormous.

For some developing countries large increases in plantation establishment are currently not feasible simply because they do not possess enough foresters to manage existing forests (Grainger, 1986). A programme to increase their planting rates significantly would mean restructuring existing government forestry departments, and the private sector if it were to be involved, in order for them to be able to cope with increased levels of activity. This would take considerable time (Grainger, 1991).

But perhaps the concern over planting rates is misplaced. More attention needs to be given to the quality of plantations and their continued management and sustainability.

10.6 Carbon Fixation as a Side Effect

Several commentators have suggested that afforestation activities in the tropics should never have carbon fixation as a primary or sole objective. For both local people, and possibly governments as well, carbon fixation would be too remote an objective and plantations established on such a basis would probably fail. And given their requirements to meet more immediate food and agricultural needs, developing countries are unlikely to make major investments related solely to climate change (FAO, 1990).

It would be very easy too for developed and developing countries alike to use large-scale plantations established on this single premise as an excuse for doing little or nothing to reduce carbon emissions. A more acceptable alternative would be to have production (subsistence or commercial) or environmental protection as a main purpose, with carbon fixation viewed as a positive side effect. Fearnside (1992) believes that if plantations were coupled with the other initiatives likely to emerge from the Climate Change Convention, that then they might have a positive role to play in terms of national CO_2 balances.

FAO's Expert Consultation on Forestry and Climate Change held in March 1990 came up with a similar conclusion, recommending that reforestation should represent sound policy independent of predicted climate change and produce net benefits separate from those which may ultimately arise in the climate change context.

10.7 Smaller-scale Initiatives

Purely in terms of storing carbon, large-scale plantations would appear to be more appropriate than agroforestry and community and social forestry initiatives; the latter tend to be small-scale and require longer to become established. However, these smaller-scale initiatives are more likely to remain viable over the long term because of their greater capacity to meet the needs of local people and the possibility of combining them with existing uses of degraded lands by rural people.

10.8 The Cost

The cost of large-scale afforestation/reforestation strategies would be high, although it is hard to determine precise figures since planting costs tend to be so generalized. DHV Consultants

Table 8. International forestry options (tropics)

	No. ha required ($x10^6$)	Planting cost/ha (US$)	Total planting cost ($x10^6$US$)	Fixation rate (tC/ha)	Carbon offset goal (Pg C/sequestered)	Planting cost/Pg carbon ($x10^9$US$)
Plant trees to offset new US electric plant CO2 (1987–96) in LDCs	6.0	400	2,400	7.5	0.045 Pg C/yr	6
Plant trees in Guatemala to offset CO_2 emissions of 180 MW electric plant in US	0.1		14[1]		0.16 PgC total over 40-year period	0.9
Restore degraded lands and provide fuelwood for year 2000 and reduce tropical C release by 40%	110	400	44,000	7.5	0.7 Pg C/yr over 40 years	6
Reforest tropical lands to offset global annual net increase in C	465	400	186,000	6.2	2.9 Pg C/yr	8
Replant tropical forest cut or burned each year	11.3/yr	400	4,500/yr		1.0 Pg C/yr	4.5

Note: [1] This number is a total project cost including forest protection, plantation establishment, and agroforestry.

Source: Andrasko (1990b)

(1991) calculated that a net increase in tropical forest cover of 12 million ha/yr would require an investment of US$18,000 million/yr or an approximately nineteen-fold increase in current annual investments in the tropical forest sector. Where would this money come from? Even if the money was found, and the meagre financial commitments made at the Rio Earth Summit make it appear unlikely, the absorptive capacity of developing country implementing institutions would pose a problem. It is already difficult for many donors to find institutions capable of running appropriate forestry projects and programmes (van Ginneken, 1991).

10.9 Unanswered Questions

Other questions currently remain unanswered in relation to large-scale afforestation. For example, what effects would the timber produced have on the existing trade, given that consumption of wood products is expected to increase at a rate of only 1.9% until the year 2000 at least, and that there may even be surpluses around 2040 (Gauthier, 1991)? Lower timber prices would certainly decrease the amount of money that could be clawed back in compensatory sales (Barnett, 1992). They could even mean that combating some problems such as those relating to pests and diseases would cease to be financially worthwhile, leading to the abandonment of plantations. Indeed, owners of existing forests might be inclined to clear their forest for more profitable land uses, thus defeating the purpose of planting trees for carbon fixation. Lower timber prices would probably also result in lower recycling rates and ultimately more waste. Much more needs to be known about the interactions between plantations and natural forests before it will be possible to make rational decisions about a reasonable allocation of resources between them (Poore in Poore *et al.*, 1989).

The effect of climate change on the plantations themselves, if they were to be established, is not precisely known. Impacts could be both positive and negative. Warmer temperatures, longer growing seasons and an atmosphere 'enriched' with CO_2, could result in increased growth. But reduced soil moisture could stress trees, making them more susceptible to insect attack and disease. Longer growing seasons might enable destructive insects to produce an additional generation, and the virulence of pathogens could increase. Drier conditions could also increase the risk of wildfire. It would therefore be crucial to match species and genotypes with site in order to minimize stress, and necessary to intensify forest protection programmes aimed at reducing losses due to fire, insects and disease.

Another area of uncertainty concerns the carbon storage potential of different soils. This must be clarified if estimates of the amounts of carbon that could be sequestered by various planting schemes are to become more accurate.

10.10 Forest Conservation

On account of some of the aforementioned difficulties, forest conservation rather than reforestation or afforestation might be a more effective forestry strategy (and more cost effective on a dollar for dollar basis) for limiting atmospheric CO_2. This is particularly so if we consider that for the late 1980s, deforestation is estimated to have accounted for between 20 and 35% of global carbon dioxide emissions (Houghton, 1990b), that forests are a reservoir for 75% of the world's carbon stored in vegetation, and that forest soils store between 55 and 73% of all soil carbon (Bolin and Johnson, 1989, in Andrasko, 1990a; Wiersum and Ketner, 1989).

Moreover, strategies to manage and maintain existing forests have the advantage of immediate beneficial impacts as against the longer term benefits of new forestation. So at the very least, they should be implemented in tandem with reforestation/afforestation. There is also potential for increasing biomass in existing poorly stocked forests (Sabhasri, 1991). Additional advantages of natural forest management include lower investment costs, because there are no land clearing costs for example, and greater adaptation to local conditions. It may be easier too, to encourage local people to become involved in natural forest management—if local knowledge and techniques are available—than to establish and maintain plantations (Mercer and Soussan, 1992).

In short, there are many variables, which differ from country to country, which need to be considered and addressed, if estimates of and plans for forestry's contribution to combating climate change are to be made (Trexler, 1991).

11. SOCIAL AND COMMUNITY FORESTRY

11.1 Fuelwood and Large-scale Plantations

Whereas industrial wood supplies are likely to be sufficient to meet demand until the end of the century and beyond, local fuelwood shortages already exist in many parts of the tropics and sub-tropics (Mather, 1990). This has not necessarily been reflected in price, though, since much fuelwood does not reach the market-place. The introduction and expansion of activities such as tea drying, brick, pottery and tile baking, based on wood as a fuel, and the production of electricity using steam from wood-fired boilers, are also putting pressure on what is a very limited resource in some areas. Plantations have appeared to offer a means of boosting fuelwood supplies. And as Goodman (1987) states, nothing has emerged so far which renders any future tree planting irrelevant, although the success of many early initiatives was questionable.

However, modern, large-scale plantation forestry, with the exception perhaps of some peri-urban fuelwood plantations (see Section 3.3), has often proved incapable of solving fuelwood problems. This can be attributed to several factors, including ill-adapted technical packages, high investment and recurrent costs, underpriced or low-valued end-products, lack of incentives to encourage local participation, underestimation of implementation difficulties and overestimation of tree growth and other benefits (Mercer and Soussan, 1992, referring to work by Floor, and Freemand and Resch). We may also add the fact that many plantations established during the 1970s when oil prices rose, became uneconomic as oil prices fell; admittedly, this reflected a policy failure in that wood from natural forests remained virtually free and hence 'cheaper'. Yet another factor is that many rural people simply do not perceive a need for fuelwood plantations.

11.2 New Perspectives

Since the late 1970s and early 1980s, attention has therefore turned instead to social forestry, community forestry, farm forestry and agroforestry, as means of solving fuelwood problems and others relating to agricultural production and rehabilitation of degraded land.

For the sake of brevity, only social and community forestry are considered here. These incorporate elements of modern, plantation forestry but differ from it in several respects. Areas planted are individually often quite small, many people may be involved, each responsible for a small part of the total overall work, and those who benefit are also usually those who carry out the planting. A further difference is that modern plantation forestry concentrates on maximizing stem volume while social and community forestry, if focusing on fuelwood production, seek to maximize branch wood volume. Social and community forestry plantations may also serve as a source of construction timber, poles, windbreaks and shelter and/or non-timber products (see also sections 3.5, 3.9 and 3.10). Community forestry also deals with managing existing forests and regeneration of degraded sites through interventions such as protection from fire and grazing.

The advantages of social and community forestry include their relatively low costs, making a significant difference to investment or financing requirements, and higher economic rates of return as a result of lower infrastructure and overhead costs. In studies of India, Kenya and the Philippines, the World Bank calculated economic rates of return of 25–30% for farm forestry as opposed to 10–15% for large-scale plantations (Spears, 1983). Leach and Mearns (1988), discussing the establishment of peri-urban plantations, state that after comparing low technology approaches, which are characteristic of social and community forestry, with high technology ones, the economic advantages of the former appear so great that choosing them becomes almost inevitable.

11.3 Problems with Social and Community Forestry

However, numerous problems have been encountered with social and community forestry. Many developing countries are in the process of shifting to capitalistic modes of production in rural areas. As a result, land has been privatized and the concept of community spirit almost made redundant. In many areas, therefore, the willingness of rural communities to become involved in establishing community plantations has been minimal (O'Keefe, 1981). Elsewhere, for example in Tanzania, tribal affiliations have hindered common action for forestry which is regarded as a non-traditional activity (Noronha, 1981). Weak community organization, for example in recently settled areas, can also be a problem.

Moreover, many social and community forestry programmes appear to have exacerbated existing inequality. Shiva *et al.* (1985) discuss World Bank social forestry programmes in Karnataka, involving afforestation with *Eucalyptus*. They consider that the benefits of these have gone to those who least need them. Village plantations have apparently extended to private landholdings that earlier were used to produce food crops. By selling fuelwood to industry, wealthier farmers have made money from their land without a corresponding dependence on the community. At the same time, agricultural output and employment opportunities, especially for women, have declined. Shyamsunder and Parameswarappa (1987) dispute these conclusions though, arguing that Shiva *et al.* are biased in their interpretation of the relevant data.

However, Romm's (1989) analysis of social forestry programmes in Asia supports some of Shiva *et al.*'s conclusions in a general sense. Although he does not refer specifically to Karnataka, he does discuss the shift from agricultural to tree crops by wealthier farmers with land of their own, who have planted trees instead of or together with food crops. He refers in particular to areas around cities where there are industrial markets for fuelwood, as well as construction timber, etc. Elsewhere, for instance in Burkina Faso, people have been reluctant to allow land, which could be used for agriculture, to be planted up with trees, and community forestry programmes have stalled as a result (Compaoré *et al.*, 1984).

And once plantations have been established, problems may be experienced in managing them. The establishment of fuelwood plantations or woodlots is sometimes organized by forestry departments in the expectation that these enterprises will later be fully taken over by the local community. But often this has not happened because the community in question is unable to meet financial requirements, or has no experience in managing such an enterprise, or is confused as to how benefits should be distributed. People may not even expect to receive any share of the benefits, considering the new plantation to be merely another category of government forest, despite the fact that it was conceived of as a common property resource.

Programmes have also failed because woodlots have been appropriated by the more powerful families in a village, particular groups such as the panchyats in India, or simply individuals. Alternatively, commons land has been regarded by foresters as 'degraded' land and used for planting woodlots, when in fact it provided grazing for the livestock of a village's poorest members. These people have often had no alternative but to continue using the newly planted areas for grazing. Moreover, village priorities are not necessarily the same as those of outsiders who might wish to give preference to the poor (Noronha, 1981).

Social and community forestry programmes also fail because many developing countries do not have sufficient rural infrastructure or extension staff to oversee initial implementation. In India in the 1980s, ever-increasing planting targets kept forest department staff busy with departmental plantations and left them with little time for extension work or for properly managing social forestry plantations. Commitment to social forestry among forestry department staff may even be low to start with. In India, foresters have often deliberately conveyed the impression that plantation management is complex and not within the capabilities of local people (Chambers *et al.*, 1989). There has also been a tendency in India for technology appropriate to large-scale plantations, with which foresters are familiar, to be applied inappropriately to village woodlots (Arnold and Stewart, 1991) and for people's species preferences to be ignored. Forest departments have favoured single-species stands as opposed to stands comprising multipurpose species or a mix of coppicing, timber, fruit and fodder trees, with the result that people have shown little interest in managing them. Likewise in the Sahel, the block-planting methods used to establish many plantations have been alien to local patterns of land use, as have the elaborate, and expensive systems used for tree protection (Taylor II and Soumaré, 1984).

For Burkina Faso, Compaoré *et al.* (1984) report that villagers were unable to assume responsibility for managing forestry and environmental conservation activities since forestry officers did not impart sufficient technical knowledge to them. This was partly because the officers themselves failed to understand why it was important that they should do so. In fact, NGOs are often institutionally much more able to work directly with local communities. But it is perhaps not surprising that forest departments originally charged with conservation and protection of state forests have found it difficult to adapt to the practices of social and community forestry. The former involved keeping people at a distance, the latter seeks their keen involvement. As Adlard (1992a) notes, there is a need to develop institutional structures which can reconcile protection of natural resources with local participation in their management.

Fuelwood-oriented programmes have also been unsuccessful because of differing perceptions. A government or forestry department may foresee an impending fuelwood scarcity. Local people, on the other hand, might not see a need to look so far ahead. Local, open-access forests may be overexploited and soil fertility declining as a result of using crop residues or dung as a substitute for fuelwood, but as far as people living in the area are concerned, fuelwood supply is not a problem. Other issues such as irrigation or road construction may receive a much higher priority (Noronha, 1981).

In many instances, therefore, the price of fuelwood will be low and other land-use alternatives such as agriculture, the returns of which come considerably more quickly than those in forestry, or growing trees for higher value products such as poles, pulpwood or lumber, will be favoured instead. As a result, it is now widely recognized that social forestry schemes should have a broader focus and seek to provide not only fuelwood but also other necessities such as poles, fodder, food, and farming materials.

Clearly, many of the problems referred to here stem from inadequate consultation with local people and a lack of appreciation of cultural and social and religious perceptions of forests, on the part of planners and foresters alike.

11.4 Reasons for Success

Cernea (1989), Noronha (1981), Spears (1983), and others, have cited the following factors as contributing to the success of social and community forestry programmes:

- strong political commitment to reforestation, including willingness of the official sector to allow local people to act as stewards of their own environment, and allocation of the necessary human, technical and financial resources;
- a tradition of communal self-help response to implementation of rural development projects;
- a certain amount of social engineering, such as group identification/formation and maintenance, setting up of lines of authority and communication channels;
- availability of sufficient land; people must not feel that reserving land for forestry will affect their other needs, particularly for grazing areas; additionally, the area set aside for forestry should be sufficiently large for harvest(s) to meet the forest product requirements of all those living in the vicinity;
- systematic use of sociological surveys to determine people's attitudes to tree planting, preferred species and other critical issues, at the outset of project development;
- the positive response of small farmers or communities to the commercial incentive of rising prices, particularly for poles and fuelwood, and the perception that farming fast growing, short rotation trees as a cash crop can be as profitable as farming some of the traditional agricultural cash crops; this factor can also have negative effects though;
- the assurance of proven technical packages for rural afforestation, which include, wherever possible, a wide range of multipurpose native and exotic trees, and which match the silvicultural technologies they promote with the social groups they aim to involve;
- demand for the products of the plantation.

DHV Consultants (1991) and Evans (1992) refer to the additional factors of:

- security of the target group over land and tree tenure;
- participation of the target group in design and management of interventions;
- participation and involvement of not simply local people generally, but local leaders and women specifically;
- use of integrated land-use planning so that all needs are met—be these for fuel, posts and poles, food, fodder, etc.

12. LAND ISSUES

12.1 Large-scale Plantations

Large-scale plantations are likely to be sited on the most suitable land if land use policies have been determined carefully on a regional, state and national basis. However, developing land use policies is a complex process involving many disciplines and considerable information gathering. For some countries, particularly in the tropics and sub-tropics, the necessary resources are simply not yet available. But even in countries which have the necessary expertise, conflict of interests, for example between forestry and agriculture or between plantation forestry and conservation, can make it extremely difficult to produce a land use policy which takes full account of land use capability and accommodates all the interests involved.

Nevertheless, a plantation project should not be implemented in isolation since it will affect its surroundings and be affected by them. Ideally then, resource, environment, economic and social assessments will be carried out to ascertain whether a plantation is viable and appropriate. Many methodologies have been assembled. Moreover, after a plantation has been established, monitoring and evaluation of its impacts should also be undertaken. Special attention needs to be given too to consulting local people on what may constitute a radical change in land use for an area.

12.1.1 Land Ownership and Tenure

In developed countries, land ownership is generally clearly defined legally. In developing countries this is often not so. Both traditional and government land tenure systems may operate, as when customary, tribal or communal ownership are not recognized under state laws, but land is nevertheless treated and used as if they are. Alternatively, customary laws and institutions may have grown weak, while state laws are not implemented because of remoteness from the influence of central government (Davidson, 1987). Plantation development may well expose tenurial ambiguities and open up conflicts which previously had remained hidden (Bass, 1992).

Even if land is privately owned, it may be used by people with no right to it. In Swaziland, expansion of the Usutu Forest in the mid-1970s had to deal with a squatting problem. Land proposed for planting had been purchased from absentee farmers but was occupied by Swazi farm-dwellers. The only solution was for the government to resettle the farmers on some of the land originally intended for plantation establishment (Evans, 1992).

If ownership is not clarified and enforced, encroachment can be a serious problem. In countries such as Brazil and Chile, where large-scale industrial plantations have been established, guarantee of legal ownership of land has therefore been a prerequisite for private investment. Thus in Chile, the government declared that land of "forest aptitude" cannot be nationalized (Binkley and Vincent, 1992). The Jarí plantations in Brazil are an exception though. Only about half of the land in the estate has had legal title confirmed, although apparently all the land already under plantations has legal title. There is also some dispute over part of the area proposed for expansion.

Corporations and investors also often consider that they should own the land to be used for plantation development because of the high level of investment in both time and money that will be required. However, outright ownership of land is not always appropriate, particularly if there are concerns over sovereignty or if existing land tenure systems do not recognize concepts of freehold. In such cases, leases and joint venture arrangements can be considered instead. Leasing means that plantation developers are made more accountable to local communities for what happens to or on the plantation land, while under joint venture arrangements, local communities can share management decisions or be compensated for giving up user rights (Bass, 1992).

If land ownership is very fragmented, acquiring sufficient land for a large-scale project can take a long time. On occasion this problem has been solved by establishing co-operatives whereby many small landowners grow the species necessary to supply an industry, as demonstrated by the PICOP project in the southern Philippines (Evans, 1992).

12.1.2 Land Use

Large-scale plantation development has sometimes failed to acknowledge, accommodate or compensate current land users. In developing countries plantations have been established on so-called unused common ground causing hardship to those least able to push their claims. Others have displaced inhabitants from their traditional hunting grounds or grazing areas, or ignored traditions, customs and beliefs about tracts of land considered to be sacred or inhabited by spirits. The Master Plans for Forest Development of several developing countries, such as Indonesia, are an obvious example. Many of these promote plantations of fast-growing commercial species without considering the dependence of local peoples on the land proposed for planting. Similarly, in Thailand, deficient land use policies meant that forest reserve land targeted for tree planting by Shell Companies of Thailand was not actually free. In effect, the Thai government left it to Shell to resolve local conflicts over land (van Ginneken, 1991; Sargent, 1990). With hindsight, it appears that large-scale plantation forestry may be most appropriate in industrial societies for whom land resources are not scarce (Kanowski *et al.*, 1992).

Even a plantation established to protect or improve resources can encounter 'opposition' if previous land use has not been taken into account. In Ethiopia, state forests were established on the basis of environmental needs as perceived by the government, and laid out by labour mobilized through food-for-work programmes. But the plantations were situated on areas subject to high human and cattle population pressures, and people continued to come into them in search of fodder and fuelwood. By 1989 it was estimated that only 7% of the seedlings planted had survived (DHV Consultants, 1991).

One of the major land use problems in less developed countries is the multiplicity of bodies dealing with land use issues. Departments of forestry, agriculture, water management and so on, may all be involved. Consequently, decision making concerning tree planting is often delayed or side-lined because of uncertainty regarding who is responsible for what. Setting up a multidisciplinary grouping, such as a department of land management, can sometimes help overcome such problems.

But even in developed countries, where land use planning is considerably more advanced, large-scale plantation establishment has sometimes been inappropriate, with the interests of local people and the context into which plantations were to be placed, being ignored. For instance, in the Basque Country, plantation establishment on common lands has left hill farmers without resources for their livestock (Groome, 1991). In fact, this has been the case throughout the Iberian Peninsula. Similarly, in Northern Spain, large-scale *Eucalyptus* plantations have been established at the expense of a decaying, archaic agricultural sector with the encouragement of the government and the industrial forestry sector. No attempt has been made to integrate the plantations, be this socially, economically or environmentally (Bernáldez *et al.*, 1989). And in the United Kingdom, planning procedures for plantations have been criticized for not encouraging or allowing sufficient participation from the public.

Complaints are also often heard from conservationists and environmentalists about the way in which large-scale plantation development has failed to take account of wildlife values. "Indicative forest strategies" are now used in the United Kingdom to attempt to reconcile any differences of opinion over proposed planting schemes.

12.1.3 Landscape Values

Landscape values and large-scale plantations are sometimes incompatible. Economies of scale in the pulp and paper industry, for instance, necessitate very large plantations (often 100,000 ha or more) if a modern mill of economic size is to be maintained. If the plantations for such a mill comprise fast-growing exotics planted in straight lines, a previously visually harmonious landscape can be destroyed. In some European countries, therefore, industrial plantations are sited away from populated areas.

Alternatively, attempts may be made to integrate the plantation into the landscape by paying greater attention to existing landscape values. Artificially straight compartment boundaries can be avoided, compartments shaped to complement individual landscape features, a mixture of textures and foliage colours provided, broadleaves planted along the boundaries of a conifer plantation, and so on. However, if the land was previously unforested, such efforts may be considered ineffective. Concern has been expressed in South Africa, for instance, over the commercial afforestation of privately owned agricultural lands in the Drakensberg Approaches. The landscape of the Approaches has a visually attractive rural character, due largely to the use of traditional agricultural methods. Much of it consists of extensive species-rich grasslands managed for grazing. Afforestation, which has already started, could easily upset the high degree of complementarity between this landscape and that of the Drakensberg Mountain Park close by, in addition to destroying natural wildlife communities (Bainbridge, 1990).

12.2 Small-scale Plantations

Small-scale plantations, for example, those established through social and community forestry programmes can fail if insufficient attention is paid to land issues. Just as with large-scale plantations, assessments have to be carried out to ascertain whether the proposed planting is appropriate to the site in question and can meet people's needs. Additionally, sufficient time and care must be devoted to consultation with local people who will be involved with or affected by the project.

12.2.1. Land Ownership and Tenure

In some countries, such as in Latin America, current land ownership laws encourage clearfelling. This can lead to plantation development (but discourage natural forest management). Elsewhere, however, it is necessary to apply for a legal permit to cut down a tree, even if it is growing on private land, and this can serve to prevent local people from participating in tree planting. Therefore, when social or community forestry programmes are being planned, legal provisions applying to tree planting and felling should be examined.

13. ECONOMIC ISSUES

The financial and economic aspects of plantation forestry are too numerous to discuss in any depth here. Therefore only key issues are presented.

13.1 Distinguishing Between Financial and Economic Assessments

Financial assessment and economic assessment are very different. Financial assessment is a limited process, generally understood as referring to the calculations a single individual or company makes in determining whether an activity would be profitable or not profitable. Economic assessment is much broader and has been equated to social cost-benefit analysis (Leslie, 1987a).

In assessing a proposed forestry activity, an economic assessment would include non-commercial benefits, commercial benefits which are difficult to express in financial terms (such as a high diversity of non-timber forest products marketed in the informal sector), and disbenefits or externalities. A financial assessment would not cover any of these. It can be particularly difficult to undertake an economic assessment, since it has to take so many factors into account and because what can be claimed as benefits by one group might be claimed by another as costs, and vice versa.

In fact, the techniques used are not yet sufficiently precise. Simply extending those developed in microeconomic theory for assessing investment by private firms, to assessment of social investments, is inadequate since the two types of investment are not strictly comparable. Investment by private firms tends to aim for monetary returns within a short timespan for a chosen set of individuals or group, whereas social investment seeks returns, or a 'good' for society at large and may involve considerably longer timespans. Even so, economic assessments should still be attempted—perhaps in combination with careful negotiation between interest groups—since they allow greater scope for identifying the potentially negative and positive impacts of proposed activities.

Clearly, financial feasibility, which in forestry often includes various incentives, has been one reason why establishing plantations has frequently been preferred to natural forest management. If more economic feasibility analyses had been carried out, many plantations would have appeared less attractive and fewer forests cleared to make way for them (Leslie, 1987b).

If a plantation is subject to purely financial analyses, severe regimes which incorporate drastic thinnings and short rotations are more likely to be favoured (Shepherd, 1986). Good management techniques may also be neglected since the returns of good management, such as maintenance of soil productivity or continuing existence of wildlife, are not assigned an immediate or sufficiently high financial value, or even any financial value at all. And if timber from natural forests is available at low prices—often evidence of inadequate valuation of forest resources—it will be harvested in preference to establishing and maintaining plantations. This has been the case in Ghana (Binkley and Vincent, 1992). Indeed, there is probably a greater need for convincing data on the true cost of forestry malpractice than for research on new forestry techniques (Ball, 1992).

13.2 Discount Rate

The discount rate enables the differences in times of occurrence of various benefits and costs to be brought back to one point in time so that a single overall cost or benefit can be calculated. But what is the best, the most appropriate rate to use? The answer may differ according to whether the calculations relate to an individual or organization, or some other subunit, or to society at large. If society is taken as the relevant unit, the [social discount] rate can be expected to be lower since it can be argued that benefits accruing to future generations are as valuable as benefits accruing to present generations and so should not be discounted.

If a high discount rate is used for comparing proposed timber production projects, plantation forestry appears preferable because of its generally shorter rotations. But if a low discount rate is used, establishing plantations may be considerably less attractive than natural forest management using longer rotations and producing a high-value end product. Natural forest also has the additional advantage of a fuller range of environmental values than most plantations. Use of high discount rates does not take this into account.

What is surprising is that the discount rate, which is a 'relative' tool, is frequently used to defend 'absolute' decisions, with no further consideration thereafter.

13.3 Sustained Finance and Risk

For large-scale plantations, sustained finance (including reasonably constant financing conditions) is crucial. Lack of finance, which is more likely to be a problem in developing countries, can mean that important activities such as site monitoring are neglected and management operations not carried out at optimal times, leading to increased chances of plantation failure. In areas where capital for plantation development is limited or uncertain, it may be more appropriate to put what money is available into natural forest management or small-scale plantations. (See also citation by Rietbergen (in Poore *et al.*, 1989) of a comparison by CTFT of the economics of plantations and natural forest management for timber production in Africa.)

Evans (1992) points out that the minimum investment life of a large-scale plantation is about 20 years, whereas the average tenure of a regime in many developing countries is often considerably less. This can lead to institutional weaknesses and, ultimately, lack of follow-up management. Understandably, this has discouraged private forestry investment in these countries (although not suggestions to establish large-scale plantations in the tropics for carbon fixation). Based on this, some commentators have argued that the biggest plantation projects should continue to be sited in temperate and industrialized countries under the control of or in collaboration with private enterprise (Gauthier, 1991). In support of this it can be argued that development costs will probably be lower in developed countries since much of the necessary infrastructure is already in place (Sedjo, 1984).

Risk is therefore not only biological, but also political and economic. Economic risk also includes fluctuations in timber and timber product prices. These make financial calculations of future returns uncertain.

13.4 Incentives

Plantation forestry is perceived as a long-term investment, and site preparation and establishment costs are frequently high. Many countries have therefore implemented financial incentive programmes, often in combination with regulations and sometimes with direct public investment, to encourage large-scale plantation establishment by private individuals or owners (McGaughey and Gregersen, 1988). In Uruguay, for example, the annual average area of plantation establishment was 2,000ha until an incentive scheme was brought in; it then shot up to 10,000ha in 1990 and 20,000ha in 1991 (Ball, 1992).

Incentives take many forms, including grants, fiscal measures, tax relief on the purchase of capital equipment, easy credit, provision of infrastructure such as roads and harbour facilities, subsidized or free inputs, and technical advice. They are justified if society will get more out of the private action than the private entity undertaking it (Gregersen *et al.*, 1979). But complex issues are involved.

In private forestry forest owners may be farmers, investors, forest products/processing companies or municipalities. Consideration should therefore be given to the types of incentives and disincentives that each group will respond to (Byron and Boutland, 1987). This has not always occurred and only one or two groups favoured instead. In 1985, after ten years of government subsidies for forest plantations, it was calculated that just three Chilean corporations had received 70% of the planting grants made available by the Chilean government (CODEFF, 1991, in Bass, 1992). Distribution problems such as these may arise because of efficiency considerations. For example, economies of scale are far easier to achieve when dealing with a few large landowners only. However, they normally have readier access to resources in the first place and are less in need of assistance than smaller landowners (Gregersen, 1984).

Moreover, incentive schemes have often encouraged inappropriate plantation establishment and misuse of natural resources. In Brazil, early incentive schemes had to be adjusted when it was found that plantations were being established on agricultural soils, on areas which were first cleared of natural forest, in remote areas with no thought given to harvesting or transporting difficulties or markets, and in such small and isolated parcels that economics of scale could not be realized (Beattie and Ferreira, 1979, in McGaughey and Gregersen, 1988). In fact, some incentive schemes have enabled individuals or groups to realize considerable profits quickly through forestry activities, at the expense of society as a whole. Thus fiscal incentives for reforestation in Brazil appear to have reflected the immediate interest of institutional structures of power and influence which aim to maintain the status quo, and have not been an engine for social change and development in the rural sector as some might have hoped (Iusem *et al.*, 1988).

Other incentive schemes have been devised in such a way that when discontinued, the plantations established through them, ceased to be maintained. This is particularly likely to occur when those taking advantage of the incentive schemes are really only interested in financial returns and not forestry *per se*.

Regulation and monitoring of an incentive scheme should therefore be carried out to evaluate not only its impacts on environmental services, but also to ascertain who is benefitting from it most. If monitoring is implemented sufficiently early it may be possible to modify a scheme if benefits are skewed or unintended environmental impacts occurring. Rapid growth of private sector forestry in a country such as Chile can be problematic if this does not take place (Hurditch, 1992).

Large-scale planting programmes executed by the private sector as a result of incentives funded by public money are probably best monitored, in terms of planting results, by the staff of national forest services (Ball, 1992). However, these people must be well trained and supported and highly motivated, otherwise the benefits of using the private sector, such as lower costs and greater flexibility, may be lost through inadequate supervision or corruption.

Incentives, both fiscal and non-fiscal, are also used to encourage small-scale plantation establishment by local people. They may be especially necessary for subsistence farmers, who by planting trees have changed the pattern of their land use and temporarily forfeited the means of providing for their daily needs. As FAO (1980, in McGaughey and Gregersen, 1988) points out, if methods of improved resource management are being introduced, those with scarce resources can hardly be expected to bear all the initial investment costs. And as with large-scale plantation establishment, the cessation of incentive schemes should not mean that the trees planted through them are neglected or removed. However, incentives for farmers to plant trees, in the form of free seedlings, for instance, have often primarily benefitted those who own land already or who can afford the space to plant trees on their land (and disbenefitted small, private nurseries at the same time).

13.5 Markets

Large-scale industrial plantation programmes must be linked to markets or industrial bases if they are to succeed financially. A country establishing extensive industrial plantations must therefore possess a large or well developed economy if it is to absorb the products grown. Export markets, whether for timber or timber products, can only be considered as an outlet if communications and transport infrastructure are adequate (Gauthier, 1991), and the timber or product highly competitive in terms of price (Douglas, 1983). In developing countries many large-scale plantations supply industries which, although established with considerable optimism in the 1950s and 1960s, have proved uncompetitive on world markets. Both merely absorb scarce capital and resources which would have been better employed in smaller-scale enterprises and which would have employed more people (Douglas, 1983).

Even if markets do exist, the long-term nature of forestry is such that by the time the product is ready for harvesting, market species or product requirements may have changed. When the Shiselweni Forestry Company in Swaziland began planting in the mid-1960s the aim was to produce mining timber from *Eucalyptus*. But by the time the crop was ready to be harvested, there was a glut of this material (Evans, 1992) and alternative markets had to be found.

If small-scale plantations are planned, adequate market studies should still be carried out if it is intended to sell the end product. Questions concerning the absorptive capacity of the market and the possibly disruptive effects of the stimulus of a cash economy, must be answered. In Bangladesh, for instance, fuelwood schemes directed towards market supply have tended to deliver maximum benefits to richer groups and minimal or zero benefits to the poorer (Douglas, 1983).

13.6 Employment

Plantations are often cited as a means of creating employment and promoting development, particularly in view of their linkages with processing industries. However, as with plantation management, each site must be treated individually with careful consideration of existing land use and markets. If an area is used for pastoral activities, the establishment of plantations could result in greater employment opportunities, although probably not for the pastoralists. But if intensive agricultural activities are pursued the opposite could be true. In areas with a growing rural population, densely planted industrial timber plantations may be unsuitable because they will offer fewer employment opportunities than many other enterprises (van Ginneken, 1991).

In Portugal, tax breaks and EC subsidies have encouraged the establishment of *Eucalyptus* plantations. However, the employment densities of these plantations are lower than those of

Table 9. Forestry incentives in Latin American countries

	DIRECT SUBSIDES						TAX BENEFITS							
							Land Use			Invested Capital				Exploitation and Forest Profits
	Credit Concession	Free Seedlings	Donation of Land and Small Planta-tions	Free Technical Assistance	Capital Donation	Cooperation in Building of Infrastructure	Exemption of all tax	Exemption of Property Tax	Exemption of Income Tax	Exemption of Inheritance and Donation Tax	Deduction in Income Tax	Exemption of all Tax	Equipment Imports Exempt or Reduced Duties	Deduction of Income Tax
Argentina	X	X				X					X			
Bolivia	X			X			X					X	X	X
Brazil	X										X			X
Chile	X	1	2	X	X			X	X	X				X (65%)
Colombia							X					X	X	
CostaRica	X	X		X				X			X			
Dominican Republic	X				X									
Ecuador							X		X				3	
El Salvador	X			X	X			X						
Guatemala	X		X					X			X		X	
Honduras														
Jamaica		X		X	X									
Mexico	X	X						X						
Nicaragua	X													
Panama		1			X		X				X		X	
Paraguay				X				X					X	
Peru	X		X		X								X	
Uruguay	X												X	
Venezuela	X			X	X									

[1] Reduced price; [2] Inexpropriability; [3] 80% reduction

Source: McGaughey and Gregersen (1988)

previous land uses involving olive groves and vineyards. The creation of 5,000ha of plantations resulted in the loss of 2,000 man-years of employment (Kardell *et al.*, 1986) and contributed to rural depopulation. In Natal/Kwazulu in South Africa, surveys have established that in areas marked for afforestation, farmers employ around one labourer for every 20 to 25ha of land. The projected level of employment by a forestry company is one labourer per 35ha. In this case, though, the lower level of employment could be offset by better conditions of employment within the forestry industry (Fourie, 1990). Edwards (1990) makes the point that it is not only numbers of jobs which are important but also improved income levels. The latter have a multiplier effect (although so too do agriculture and livestock) and increase the circulation of money, enabling further wealth to be created.

The end uses for which plantation products are destined should also be considered. The capital-intensive pulp and paper industry creates little direct employment and adds relatively little value to the timber processed, in contrast to an industry such as furniture making (Bernáldez *et al.*, 1989). In countries where unemployment is high, such issues merit careful consideration before plantations are established.

Compromises, for example, between the need for a large supply of pulp and the need to create rural employment can be worked out. Bernáldez *et al.* (1989) suggest that in northern Spain, improved management of *Eucalyptus* plantations could increase the employment opportunities offered by the plantations and double their yield. Pulpwood supplies would be maintained but the area required for *Eucalyptus* plantations reduced. The 'excess' land could then be used to grow high quality timber for use in small-scale rural industries such as furniture making, and create further employment and wealth. However, such a scheme would require financial incentives in order to make it a viable financial proposition.

14. CONCLUSIONS

As has been stated elsewhere in this report, plantations are established for many different reasons and under many different conditions. This makes generalizing about them difficult and sometimes even unhelpful. However, there are some guiding principles or statements which IUCN considers to be broadly applicable.

Definitions and data:

- Plantation types, or what is regarded as a plantation, should be defined more clearly, and the definitions then applied universally so that figures can be compared. There is also a need for more information on plantation survival rates in order to improve the estimates of plantation extent.

- It would be useful to have a standard definition for the term "degraded", in order that scenarios which refer to planting up degraded land could be compared and to ensure that existing uses of poor quality land by rural poor are not discounted.

Current deforestation, reforestation, afforestation:

- Currently, most reforestation and afforestation takes place in the temperate and sub-tropical zones, whereas deforestation is concentrated in low latitudes. Forestry is perhaps not focusing sufficiently on areas where it is particularly needed.

Justification:

- Given the variety of purposes they can fulfil, there is a need for both large- and small-scale plantations.

Institutional Arrangements:

- Plantation forestry should be placed within a broader context of rural development and land use.

- There is a need to integrate the various agencies who deal with land use, so that both large- and small-scale plantation forestry can be promoted and regulated effectively.

- Institutions able to integrate the complex aspects of forestry within the broader context of development and management of rural landscapes should be established. Particular attention should be paid to setting up institutions which incorporate different professional backgrounds and expertise so as to create true interdisciplinary teams.

Land use:

- Plantations and natural forests are complementary rather than alternative land uses.

- Natural or mature secondary forest should not be cleared to make way for plantations. However, all habitats should be evaluated prior to planning and implementing reforestation/afforestation activities.

- Existing land use(s) should be assessed prior to the implementation of plantation schemes. Studies to identify and evaluate potential environmental and social impacts should be undertaken to ascertain whether the proposed change in land use is justified.

Planning:

- Plantation objectives must be clearly defined and realistic if they are to be attained, and management plans must follow logically from these objectives. Large-scale plantations, in particular, are high risk enterprises and require careful planning, sustained finance and timely management if they are to be successful.

Local communities:

- Local people should be consulted prior to and during the planning, implementation and monitoring of plantation schemes. Wherever possible, they should be involved in the management of plantations. No community land should be alienated for plantation establishment without the full agreement and cooperation of local communities.

Species:

- All species should be matched carefully with site.
- Indigenous species are to be preferred for plantations if they are equal to or better than exotics for the purpose(s) specified.
- There is a need for more research into all aspects of growing, managing and reproduction of species, but particularly in relation to promoting utilization of a wider range of native species/genera, for a variety of purposes.
- Exotics should be selected with considerable care and only after it has been ascertained —as far as possible—that they will integrate well into their proposed new environment.
- Greater efforts should be made to study and conserve the sources of exotic material in their native habitat.
- Greater research into the benefits and drawbacks of mixed species plantations is required. Mixed species planting should be considered for community forestry, firebreaks, as nurse crops and so on.

Silvicultural management:

- Nutrient, water, and pest and disease problems can be largely avoided or minimized through good management and monitoring. The failure of many large-scale plantations is attributable to poor management.
- Greater efforts should be made to optimize use of mineral fertilizer and chemical pesticides. IPM should form the basis of management of forest pests.

Biodiversity:

- Plantation managers can adopt various management techniques to accommodate some wildlife needs, although incentives or legal requirements may often be necessary to persuade them to do so.
- Biodiversity conservation and small-scale plantations (particularly in relation to the tropics, where small-scale plantation establishment is developing rapidly) is an area requiring further research. There is potential for the two interests to complement each other's aims and this should be developed.
- Natural forest remnants, natural forest corridors and old growth forest, should, wherever possible, be incorporated in intensively managed plantation areas.

Climate change:

- Large-scale plantations should not be used as a 'primary tool' for carbon fixation. In terms of value for money and immediate benefits, conservation and management of existing forests, and smaller-scale initiatives, involving agroforestry for example, would be more appropriate. Carbon fixation should rather be aimed at as a beneficial side effect.
- More scenarios should be developed to examine the potential impact on timber markets of further afforestation/reforestation and of increases in CO_2 levels on plantation health and management.

Incentives:

- Government incentives for private plantation development should only be provided under contracts stipulating the social and environmental benefits to be realized in return for government support. They are justified only if society as a whole benefits more than the private sector undertaking the forestry development activity.
- Incentive schemes for forestry should be monitored and regulated to ensure that benefits are not skewed and to avoid or minimize negative environmental, social and economic impacts.

Assessment:

- Standardized means for assessing some of the environmental, social and economic impacts of plantations could be devised. However, any large-scale plantation must be subject to a full assessment of its environmental, social and economic impacts before any definitive judgement is made of its merits or demerits.
- Economic assessment as opposed to financial assessment should be undertaken for plantation schemes. Collection and development of data on the full costs of forestry mismanagement are required.

REFERENCES

Adlard, P. 1992a. *Tree Plantation Review: Study No 5: Sustainability.* Shell International Petroleum Company/WWF, UK.

Adlard, P. 1992b. *Tree Plantation Review: Study No 1: Historical Background.* Shell International Petroleum Company/WWF, UK.

Ahmed, S., Bamofleh, S. and Munshi, M. 1989. Cultivation of neem, (*Azadirachta indica*, Meliaceae) in Saudi Arabia. *Economic Botany* 43(1).

Alexander, M. 1989. The long-term effect of *Eucalyptus* plantations on tin-mine spoil and its implication for reclamation. *Landscape and Urban Planning* 17.

Andrasko, K. (Based on the work of). 1990a. Climate Change and Global Forests: Current Knowledge of Potential Effects, Adaptation and Mitigation Options. FAO, Rome. Unpublished report.

Andrasko, K. 1990b. (Revised draft). Forestry chapter. In: Lashof, D. and Tirpak, D. (Eds), *Policy Options to Stabilize Global Climate.* Report to Congress by Office of Policy Analysis, US Environmental Protection Agency.

Andrasko, K., Heaton, K. and Winnett, S. 1991. Evaluating the costs and efficiency of options to manage global forests: a cost curve approach. In: *Proceedings of the Technical Workshop to Explore Options for Global Forestry Management, Bangkok, Thailand, 24–30 April 1991.* IIED, London.

Arnold, J.E.M. and Stewart, J.C. 1991. *Common Property Resource Management in India.* Tropical Forestry Paper 24. Oxford Forestry Institute, Oxford.

Attiwill, P.M., Guthrie, H. and Leuning, R. 1978. Nutrient cycling in a *Eucalyptus obliqua forest.* IV nutrient uptake and nutrient return. *Australian Journal of Botany* 26.

Austara, O. 1971. *Gonometa podocarpi* Aur. (Lepidoptera: Lagiocampidae). A defoliator of exotic softwoods in East Africa. Biology and life cycle at Muko, Uganda. *East Africa Agriculture and Forestry Journal* 36.

Baharudin, J. 1987. An appraisal of the compensatory plantation programme in Peninsular Malaysia. In: Liew That Chim (Ed.), *Proceedings: Seminar on Forest Plantation Development in Malaysia, 30 November–4 December 1987, Kota Kinabalu, Sabah, Malayasia.* Forest Department, Sabah.

Bainbridge, W.R. 1990. Commercial afforestation in the Approaches zones of the Natal Drakensberg as a potential threat to environmental quality and the integrity of natural communities in the Drakensberg region. In: Erskine, J.M. (Ed.), The Physical, Social and Economic Impacts of Large-scale Afforestation in Natal/Kwazulu. (Unpublished) Proceedings of the Forestry Impacts Workshop, 8 May 1990. Institute of Natural Resources, University of Natal, Pietermaritzburg.

Ball, J.B. 1992. Forest plantations and the wise management of tropical forests. In Adam, K.L. and Miller, F.R. (Eds), *Proceedings of the Oxford Conference on Tropical Forests: Wise Management, 30 March–1 April 1992.* Oxford Forestry Institute, Oxford.

Barbier, E.B. 1987. The concept of sustainable economic development. *Environmental Conservation* 14(2).

Barnett, A. 1992. *The Environmental and Social Impacts of Large-scale Tropical Reforestation in Response to Global Climate Change*. Friends of the Earth, London.

Bass, S. 1992. *Tree Plantation Review: Study No 7: Social Environment*. Shell International Petroleum Company/WWF, UK.

Bernáldez, F.G., Ruiz, M. and Levassor, C. 1989. Effets des Plantations d'Eucalyptus dans le Nord de l'Espagne. EEC, Brussels. Unpublished report.

Binkley, C.S. and Vincent, J.R. 1992. Forest-based industrialization: a dynamic perspective. In: Sharma, N.P. (Ed.), *Managing the World's Forests: Looking for Balance between Conservation and Development*. Kendall/Hunt, Iowa.

Boardman, R. 1979. Maintenance of productivity in successive rotations of radiata pine in South Australia. In Ford E.D., Malcolm, D.C. and Atterson, J. (Eds), *The Ecology of Even-aged Forest Plantations*. Institute of Terrestrial Ecology, Cambridge, UK.

Brand, D.G., Kehoe, P. and Connors, M. 1986. Coniferous afforestation leads to soil acidification in central Ontario. *Canadian Journal of Forest Research* Vol 16.

Brasier, C.M. 1992. Oak tree mortality in Iberia. *Nature* Vol 360, 10 December.

Bruijnzeel, L.A. 1989. De/forestation and dry season flow in the tropics: a closer look. *Journal of Tropical Forest Science* i(3).

Budowski, C. 1984. Biological diversity and forestation in the tropics. In: Wiersum, K.F. (Ed.), *Strategies and Designs for Afforestation, Reforestation and Tree Planting: Proceedings of an International Symposium on the Occasion of 100 Years of Forestry Education and Research in the Netherlands, Wageningen, 19–23 September, 1983*. Pudoc, Wageningen.

Bull, P.C. 1981. The consequences for wildlife of expanding New Zealand's forest industry. *New Zealand Journal of Forestry* Vol 26.

Burdon, R.D. 1982. Monocultures—how vulnerable? *What's New in Forest Research?* No 115. Forest Research Institute, Rotorua, New Zealand.

Byron, N. and Boutland A. 1987. Rethinking private forestry in Australia. 1: Strategies to promote private timber production. *Australian Forestry* 50(4).

Camino Velozo, R. 1987. *Incentives for Community Involvement in Conservation Programmes*. FAO Conservation Guide No 12. FAO, Rome.

Campinhos Jr, E. 1991. Plantation of fast-growing species for tropical areas. In: *Proceedings 5, 10th World Forestry Congress*.

Carlson, A. 1986. A comparison of birds inhabiting pine plantations and indigenous forest patches in a tropical mountain area. *Biological Conservation* Vol 35 No 3.

Centeno, J.C. 1992. *The Need to Reforest the Tropics*.

Cernea, M.M. 1989. User groups as producers in participatory afforestation strategies. World Bank Discussion Paper 70. World Bank, Washington, DC.

Chambers, R., Saxena, N.C. and Shah, T. 1989. *To the Hands of the Poor*. Intermediate Technology Publications, New Delhi.

Chou, C.K.S. 1981. Monoculture, species diversification, and disease hazards in forestry. *New Zealand Journal of Forestry* Vol 26 No 1.

Chuntanaparb, L., Sri-Aran P. and Hoamuangkaew, W. 1985. *Non-wood Forest Products in Thailand. Special Study on Forest Management, Afforestation and Utilization of Forest Resources in the Developing Regions.* FAO, Bangkok.

Ciesla, W.M. 1992. Recent introductions of forest insects and their effects: a worldwide overview. Paper presented to the Regional Conference on the Management of the Woodwasp *Sirex noctilio* in South American Pine Plantations. Florianopolis, Brazil 23 –27 November 1992.

Compaoré, A., Laban, P., Visser, W. and Zongo, A. 1984. Les actions bois des villages en Haute-Volta. In: Wiersum, K.F. (Ed.), *Strategies and Designs for Afforestation, Reforestation and Tree Planting: Proceedings of an International Symposium on the Occasion of 100 Years of Forestry Education and Research in the Netherlands, Wageningen, 19–23 September, 1983.* Pudoc, Wageningen.

Cooper, K. 1990. Commercial forestry impacts on wetlands, grasslands and indigenous forests. In: Erskine, J.M. (Ed.), The Physical, Social and Economic Impacts of Large-scale Afforestation in Natal/Kwazulu. (Unpublished) Proceedings of the Forestry Impacts Workshop, 8 May 1990. Institute of Natural Resources, University of Natal, Pietermaritzburg.

DHV Consultants. January 1991. Study on the Potential Contribution of Tropical Forestry to Decrease Atmospheric Carbon Content. DHV Consultants, Amersfoort. Unpublished report.

Davidson, J. 1987. *Bioenergy Tree Plantations in the Tropics: Ecological Implications and Impacts.* Commission on Ecology Paper No 12. IUCN, Gland, Switzerland.

Lam Cong Dinh. 1992. Sylviculture et agroforesterie dans la reédification du patrimonie boise au Vietnam. In: *Proceedings 5, 10th World Forestry Congress.*

Dione, M. 1987. Rôle des peuplements d'acacia-gommier de la zone nord et nord est du Sénégal: bilan d'actions et perspectives. In: Les Arbres fixateurs d'Azote. L'Amélioration biologique de la fertilité du sol. Actes des Séminaires, 17–25 mars, Dakar. ORSTOM, Paris.

Douglas, J.J. 1983. *A Re-appraisal of Forestry Development in Developing Countries.* Martinus Nijhoff/Dr W. Junk Publishers, Dordrecht.

Duff, A.B., Hall, R. A. and Marsh, C.W. 1986. A survey of wildlife in and around a commercial tree plantation in Sabah. *Malaysian Forester* 47.

Dyck, W.J. and Beets, P.N. 1987. Managing for long-term site productivity. *New Zealand Forestry*, November.

Edwards, M.B.P. 1990. The economic significance of the forestry and forest products industry to the Natal/Kwazulu region. In: Erskine, J.M. (Ed.), The Physical, Social and Economic Impacts of Large-scale Afforestation in Natal/Kwazulu. (Unpublished) Proceedings of the Forestry Impacts Workshop, 8 May 1990. Institute of Natural Resources, University of Natal, Pietermaritzburg.

van Emden, H.F. and Williams, G.F. 1974. Insect stability and diversity in agro-ecosystems. *Annual Review of Entomology* 19.

Evans, J. 1985. A time to plant—the complementary need to expand intensive forestry programmes in developed and developing countries. *Journal of World Forest Resource Management* Vol 1.

Evans, J. 1986a. Plantation forestry in the tropics—trends and prospects. *The International Tree Crops Journal* 4.

Evans, J. 1986b. Productivity of second and third rotations of pine in the Usutu Forest, Swaziland. *Commonwealth Forestry Review* (65) 3.

Evans, J. 1987. Site and species selection—changing perspectives. *Forest Ecology and Management* No 21, 3,4, November.

Evans, J. 1988. The Usutu forest: twenty years later. *Unasylva* 40(159).

Evans, J. 1992. *Plantation Forestry in the Tropics*. 2nd edition. Clarendon Press, Oxford.

FACE (Forests Absorbing Carbon Dioxyde Emission [sic]). 1991. Booklet.

FAO. 1967. *Proceedings of the World Symposium on Man-made Forests and their Industrial Importance* Vol I. FAO, Rome.

FAO. 1990. Climate change and agriculture, forestry and fisheries. FAO position paper. FAO, Rome.

FAO, 1991. Second interim report on the state of tropical forests. Forest Resources Assessment 1990 Project. Presented at 10th World Forestry Congress, Paris, September.

Fearnside, P.M. 1983. Development alternatives in the Brazilian Amazon: an ecological evaluation. *Interciencia* 8(2).

Fearnside, P.M. 1992. Tropical silvicultural plantations as a means of sequestering atmospheric carbon dioxide. Unpublished paper.

Fearnside, P.M. (in press). Agroforestry in Brazil's Amazonian development policy: the role and limits of a potential use for degraded lands. In: Sachs, I. (Ed.), *Ecologically, Socially and Economically Sustainable Resource Use Patterns in the Humid Tropics*. UNESCO, Paris.

Forestry Commission. 1988. Forest Facts 3: Timber Production in Britain. Booklet.

Fourie, J.J. 1990. Organized agriculture's concern over large-scale afforestation in Natal/Kwazulu. In: Erskine, J.M. (Ed.), The Physical, Social and Economic Impacts of Large-scale Afforestation in Natal/Kwazulu. (Unpublished) Proceedings of the Forestry Impacts Workshop, 8 May 1990. Institute of Natural Resources, University of Natal, Pietermaritzburg.

Friend, G.R. 1980. Wildlife conservation and softwood forestry in Australia: some considerations. *Australian Forestry* 43(4).

Friend, G.R. 1982. Mammal populations in exotic pine plantations and indigenous eucalypt forests in Gippsland, Victoria. *Australian Forestry* 45(1).

Gabbrielli, A. 1990. Recent research on forest history. *News of Forest History* No13/14.

Gauthier, J.J. 1991. Les bois de plantations dans le commerce mondial des produits forestiers. In: *L'Emergence des nouveaux potentiels forestiers dans le monde*. AFOCEL, Paris.

Gibson, I.A.S. and Jones, T. 1977. Monoculture in the origin of major forest pests and diseases. In: Cherrett, J.M. and Sagar, G.R. (Eds), *Origins of Pest, Parasite, Disease and Weed Problems*. Blackwell Scientific Publications, Oxford, London, Edinburgh, Victoria.

van Ginneken, P. 1991. Not Seeing the People for the Trees. AIDEnvironment, Amsterdam. Unpublished report.

Gladstone, W.T. and Ledig F.T. 1990. Reducing pressure on natural forests through high-yield forestry. *Forest Ecology and Management* 35(1990).

Good, J., Lawson, G. and Stevens P. 1992. *Tree Plantation Review: Study No 8: Natural Environment*. Shell International Petroleum Company/WWF, UK.

Goodman, G. 1987. Biomass energy in developing countries: problems and challenges. *Ambio* Vol 16 Nos 2–3.

Grainger, A. 1986. Deforestation and progress in afforestation in Africa. *International Tree Crops Journal* 4.

Grainger, A. 1988. Estimating areas of degraded tropical lands requiring replenishment of forest cover. *International Tree Crops Journal* 5(1–2).

Grainger, A. 1990. Modelling the impact of alternative afforestation strategies to reduce carbon emissions. In: *Proceedings of the Intergovernmental Panel on Climate Change. Conference on Tropical Forestry Response Options to Global Climate Change, 9–12 January 1990*. Report No 20P-2003. Office of Policy Analysis, US Environmental Protection Agency, Washington, DC.

Grainger, A. 1991. Constraints on increasing tropical forest area to combat global climate change. In: *Proceedings of the Technical Workshop to Explore Options for Global Forestry Management, Bangkok, Thailand, 24–30 April 1991*. IIED, London.

Gray, B. 1972. Economic tropical entomology. *Annual Review of Entomology* 17.

Gregersen, H. 1984. Incentives for forestation: a comparative assessment. In: Wiersum, K.F. (Ed.), *Strategies and Designs for Afforestation, Reforestation and Tree Planting: Proceedings of an International Symposium on the Occasion of 100 Years of Forestry Education and Research in the Netherlands, Wageningen, 19–23 September, 1983*. Pudoc, Wageningen.

Gregersen, H., Houghtaling, T. and Rubinstein, A. 1979. Economics of public forestry incentive programs: a case study of cost-sharing in Minnesota. Technical Bulletin No 315. Agricultural Experiment Station, University of Minnesota.

Groome, H. 1991. Conflicts caused by imbalances in forest policy and practice in the Basque Country. *Progress in Rural Policy and Planning* Vol 1. Belhaven Press, London and New York.

Gustaffson, L. 1987. Plant conservation aspects of energy forestry—a new type of land use in Sweden. *Forest Ecology and Management* 21.

Hamilton, L.S. and King, P.N. 1983. *Tropical Forested Watersheds: Hydrologic and Soils Response to Major Uses or Conversions*. Westview Press, Boulder, Colorado.

Harriman, R. and Morrison, B.R.S. 1981. Forestry, fisheries and acid rain in Scotland. *Scottish Forestry* 35.

Hartshorn, G.S. 1983. Ecological implications of tropical plantation forestry. In: Sedjo, R. *The Comparative Economics of Plantation Forestry: A Global Assessment*. Resources for the Future, Washington, DC.

Hase, H. and Foelster, H. 1983. Impact of plantation forestry with teak (*Tectona grandis*) on the nutrient status of young alluvial soils in West Venezuela. *Forest Ecology and Management* 6 (1983).

Hill, M.O. 1979. The development of a flora in even-aged plantations. In: Ford, E.D., Malcolm, D.C. and Atterson, J. (Eds), *The Ecology of Even-aged Forest Plantations*. Institute of Terrestrial Ecology, Cambridge, UK.

van Hook, R.I., Johnson, D.W., West, D.C. and Mann, L.K. 1982. Environmental effects of harvesting forests for energy. *Forest Ecology and Management* 4.

Houghton, R. 1990a. The future role of tropical forests in affecting the carbon dioxide concentration of the atmosphere. *Ambio* Vol 19 No 4.

Houghton, R. 1990b. The global effects of tropical deforestation. *Environmental Science and Technology* 24.

Houghton, R.A. and Woodwell, G.M. 1989. Global climatic change. *Scientific American* Vol 260(4).

Hummel, F.C. 1989. Biomass forestry in Europe: strategy for the future. In: Grassi, G., Pirrwitz, D. and Zibetta, H. (Eds), *Energy from Biomass 4*. Elsevier, London and New York.

Hurditch, W.J. 1992. *Problems of Public Forestry and the Socioeconomic Implications of Privatization*. OFI Occasional Paper No 42. Oxford Forestry Institute, Oxford.

IUCN/UNEP/WWF. 1980. *World Conservation Strategy: Living Resource Conservation for Sustainable Development*. IUCN/UNEP/WWF, Gland, Switzerland.

Iusem, A.N., Correa de Lima, J.P. and Mercado, R.S. 1988. Brazil: the forest sector's participation in international trade. In: Nagy, A. (Ed.), *International Trade in Forest Products*. A.B. Academic Publishers, Oxon.

Kanowksi, P.J. and Savill, P.S. 1992. Forest plantations: towards sustainable practice. In Sargent, C. and Bass, S. (Eds), *Plantation Politics: Forest Plantations in Development*. Earthscan, London.

Kanowski, P.J. and Savill, P.S. with Adlard, P.G., Burley, J., Evans, J., Palmer, J.R. and Wood, P.J. 1992. In: Sharma, N.P. (Ed.), *Managing the World's Forests: Looking for Balance between Conservation and Development*. Kendall/Hunt, Iowa.

Kardell, L., Steen, E. and Fabiao, A. 1986. Eucalyptus in Portugal—a threat or a promise? *Ambio* 15 No 1.

Khen, C.V., Speight, M.R. and Holloway, J.D. 1992. Comparison of biodiversity between rain forest and plantations in Sabah using insects as indicators. In Adam, K.L. and Miller, F.R. (Eds), *Proceedings of the Oxford Conference on Tropical Forests: Wise Management, 30 March–1 April 1992*. Oxford Forestry Institute, Oxford.

Kimmins, J.P. 1977. Evaluation of the consequences for future tree productivity of the loss of nutrients in whole-tree harvesting. *Forest Ecology and Management* 1.

Kio, P.R.O. and Ekwebelam, S.A. 1987. Plantations versus natural forests for meeting Nigerian wood needs. In: Mergen, F. and Vincent, J.R. (Eds), *Natural Management of Tropical Moist Forests*. Yale University, Newhaven, Connecticut.

Lanly, J.P. 1982. *Tropical Forest Resources*. FAO Forestry Paper No 30. FAO, Rome.

Leach, G. and Mearns, R. 1988. *Beyond the Woodfuel Crisis*. Earthscan, London.

Lee, S.A. 1992. *Tree Plantation Review: Study No 6: Biodiversity*. Shell International Petroleum Company/WWF, UK.

Leslie, A.J. 1987a. The economic feasibility of natural management of tropical forests. In: Mergen, F. and Vincent, J.R. (Eds), *Natural Management of Tropical Moist Forests*. Yale University, Newhaven, Connecticut.

Leslie, A.J. 1987b. A second look at the economics of natural management systems in tropical mixed forests. *Unasylva* 155 Vol 39.

Logan, W.E.M. 1967. FAO World Symposium on Man-made Forests and their Industrial Importance: 1 Policy. *Unasylva* 21(66–7).

Lugo, A.E. 1992. Comparison of tropical tree plantations with secondary forests of similar age. *Ecological Monographs* 62(1).

Mather, A.S. 1990. *Global Forest Resources*. Belhaven Press, London.

Mather, A. (Ed.) 1993. *Afforestation: Policies, Planning and Progress*. Belhaven Press, London and Florida.

McGaughey, S.E. and Gregersen, H.M. 1988. *Investment Policies and Financing Mechanisms Available for Sustainable Forestry Development*. Inter-American Development Bank, Washington, DC.

Mercer, D.E. and Soussan, J. 1992. Fuelwood problems and solutions. In: Sharma, N.P. (Ed.), *Managing the World's Forests: Looking for Balance between Conservation and Development*. Kendall/Hunt, Iowa.

Miller, H.G. 1979. The nutrient budgets of even aged forests. In: Ford, E.D, Malcolm, D.C. and Atterson, J. (Eds), *The Ecology of Even-aged Forest Plantations*. Institute of Terrestrial Ecology, Cambridge, UK.

Montagnini, F. and Sancho, F. 1990. Impacts of native trees on tropical soils: a study in the Atlantic lowlands of Costa Rica. *Ambio* Vol 19 No 8.

Murray, J.S. 1979. The development of populations of pests and pathogens in even-aged plantations—fungi. In: Ford, E.D, Malcolm, D.C. and Atterson, J. (Eds), *The Ecology of Even-aged Forest Plantations*. Institute of Terrestrial Ecology, Cambridge.

Newhook, F.J. and Podger, F.D. 1972. The role of *Phytophthora cinnamomi* in Australian and New Zealand forests. *Annual Review of Phytopathology* 10.

Noronha, R. 1981. Why is it so difficult to grow fuelwood? *Unasylva* Vol 33 No 131.

Nutter, W.L. 1979. Effects of forest plantations on the quantity, quality and timing of water supplies. In: Ford, E.D, Malcolm, D.C. and Atterson, J. (Eds), *The Ecology of Even-aged Forest Plantations*. Institute of Terrestrial Ecology, Cambridge.

Nwoboshi, L.C. 1987. Regeneration success of natural management, enrichment planting and plantations of native species in West Africa. In: Mergen, F. and Vincent, J.R. (Eds), *Natural Management of Tropical Moist Forests*. Yale University, Newhaven, Connecticut.

Nyaga, C.R.J. 1991. Kenyan experiences in forest management: dry tropical forest. In: *Proceedings of the Technical Workshop to Explore Options for Global Forestry Management, Bangkok, Thailand 24–30 April 1991*. IIED, London.

O'Keefe, P. 1981. Background papers prepared by Beijer Institute for the UN Renewable Energy Conference, Nairobi.

Ormazábal, C.S. 1991. Silvopastoral systems in arid and semi-arid zones of Northern Chile. *Agroforestry Systems* 14.

Perry, D.A. and Maghembe, J.A. 1989. Ecosystem concepts and current trends in forest management: time for reappraisal. *Forest Ecology and Management* 26.

Persson, R. 1974. World forest resources: review of the world's forest resources in the early 1970s. Research Note 17. Royal College of Forestry Stockholm.

Pohjonen, V. and Pukkala, T. 1990. *Eucalyptus globulus* in Ethiopian forestry. *Forest Ecology and Management* 36.

Poore, D. 1992. *Tree Plantation Review: Study No 12: Guidelines*. Shell International Petroleum Company/WWF, UK.

Poore, D., Burgess, P., Palmer, J., Rietbergen, S. and Synott, T. 1989. *No Timber Without Trees*. Earthscan, London.

Poore, M.E.D. and Fries, C. 1985. *The Ecological Effects of Eucalyptus*. FAO Forestry Paper 59. FAO, Rome.

Porter, R. 1990. Future afforestation and the potential impacts on nature conservation in Natal. In: Erskine, J.M. (Ed.), The Physical, Social and Economic Impacts of Large-scale Afforestation in Natal/Kwazulu. (Unpublished) Proceedings of the Forestry Impacts Workshop, 8 May 1990. Institute of Natural Resources, University of Natal, Pietermaritzburg.

Postel, S. and Heise, L. 1988. *Reforesting the Earth*. Worldwatch Paper No 83. Worldwatch Institute, Washington, DC.

Pryor, L.D. 1978. Eucalypts as exotics. Inter-training course in forest tree breeding. Australian Development Assistance Agency, Canberra.

Quezel, P., Barbero, M. and Loisel, R. 1990. Les reboisements en région méditerranéenne. Incidences biologiques et économiques. *Forêt Méditerranéenne* t.XII No 2.

Ranger, J., Robert, M., Bonnaud, P. and Nys, C. 1990. Les minéraux-test, une approche expérimentale *in situ* de l'altération biologique et du fonctionnement des écosystèmes forestiers. Effets des types de sols et des essences feuillues et résineuses. *Ann Sci For* 21.

Rao, Y.S. 1991. State of forestry in the Asia-Pacific region. In: *Proceedings of the Technical Workshop to Explore Options for Global Forestry Management, Bangkok, Thailand 24–30 April 1991*. IIED, London.

Reijneveld, A. 1983. A comparison of various estimates of reafforestation areas in the tropics. BOS Document No 1. Dutch Forestry Development Co-operation.

Romm, J. 1989. Forestry for development: some lessons from Asia. *Journal of World Forest Resources* Vol 4.

le Roux, S.D. 1990. Impact of afforestation on the agricultural resources of Natal. In: Erskine, J.M. (Ed.), The Physical, Social and Economic Impacts of Large-scale Afforestation in Natal/Kwazulu. (Unpublished) Proceedings of the Forestry Impacts Workshop, 8 May 1990. Institute of Natural Resources, University of Natal, Pietermaritzburg.

Sabhasri, S. 1991. Opening speech. In: *Proceedings of the Technical Workshop to Explore Options for Global Forestry Management, Bangkok, Thailand, 24–30 April 1991*. IIED, London.

Sargent, C. 1990. The Khun Song Plantation Project: a Socioeconomic and Environmental Analysis and Recommendations Towards the Establishment and Management of a Plantation in Chanthaburi Province by Shell Companies of Thailand. IIED, London. Unpublished report.

Sargent, C. and Lowcock M. 1991. Options for the coordination of international action on forest conservation and management. In: *Proceedings of the Technical Workshop to Explore Options for Global Forestry Management, Bangkok, Thailand, 24–30 April 1991*. IIED, London.

Savill, P.S. and Evans, J. 1986. *Plantation Silviculture in Temperate Regions: With Special Reference to the British Isles*. Clarendon Press, Oxford.

Schönau, A.P.G. 1984. Silvicultural considerations for high productivity of *Eucalyptus grandis*. *Forest Ecology and Management* 9.

Sedjo, R.A. 1984. An economic assessment of industrial forest plantations. *Forest Ecology and Management* 9.

Sedjo, R.A. 1987. Forest resources of the world: forests in transition. In: Kallio, M., Dykstra, D.P. and Binkley, C.S. (Eds), *The Global Forest Sector: An Analytical Perspective*. Wiley-Interscience, New York.

Sedjo, R.A. and Clawson, M. 1984. *Global Forests*. Resources for the Future, Washington, DC.

Senanayake, F.R. 1987. Analog forestry as a conservation tool. *Tiger Paper, Forest Newsletter* 14(2).

Shell International Petroleum Company Ltd. 1990. Shell in Forestry. Shell Briefing Service Number 5. Shell Publications Limited. Booklet.

Shepherd, K.R. 1986. *Plantation Silviculture*. Martinus Nijhoff Publishers, Dordrecht.

Shiva, V., Bandyopadhyay, J. and Jayal, N.D. 1985. Afforestation in India: problems and strategies. *Ambio* 14 No 6.

Shyamsunder, S. and Parameswarappa, S. 1987. Forestry in India—the forester's view. *Ambio* Vol 16 No 6.

Simmonds, F.J. and Greathead, D.J. 1977. Introductions and pest and weed problems. In: Cherrett, J.M. and Sagar, G.R. (Eds), *Origins of Pest, Parasite, Disease and Weed Problems*. Blackwell Scientific Publications, Oxford, London, Edinburgh, Victoria.

Siren, G. 1980. Present stage of energy forestry in Sweden. In: *Proceedings of International Conference on Harvesting and Utilization of Wood for Energy Purposes, Jonkoping, Sweden*. Report No 19. Swedish University Agricultural Science, Garpenderg.

Spears, J.S. 1983. Tropical reforestation: an achievable goal? *Commonwealth Forestry Review* (62)3.

Spears, J.S. 1991. Plantation forestry in developing countries: equity, environmental and economic concerns. Discussion paper presented at Centre for Applied Studies in International Negotiations (CASIN) Symposium on Tree Plantations: Benefits and Drawbacks, Geneva, 23–24 April 1991.

Speight, M.R. and Wainhouse, D. 1989. *Ecology and Management of Forest Insects*. Oxford Science Publications, Oxford.

Squire, R. O., Farrell, P.W., Flinn, D.W. and Aeberli, B.C. 1985. Productivity of first and second rotation stands of radiata pine on sandy soils 2. Height and volume growth at five years. *Australian Forestry* 48(2).

Suchek, V.I. 1991. The role of the planted forest in the pulp and paper industry in Brazil. *The Forestry Chronicle* Vol 67 No 6.

Sutton, W. 1991. Are we too concerned about wood production? *NZ Forestry* 36(3).

Swank, W.T. and Douglas, J.E. 1974. Streamflow greatly reduced by converting deciduous hardwood stands to pine. *Science* 185.

Taylor II, G.F. and Soumaré, M. 1984. Strategies for forestry development in the semi-arid tropics: lessons from the Sahel. In: Wiersum, K.F. (Ed.), *Strategies and Designs for Afforestation, Reforestation and Tree Planting: Proceedings of an International Symposium on the Occasion of 100 Years of Forestry Education and Research in the Netherlands, Wageningen, 19–23 September, 1983*. Pudoc, Wageningen.

Trexler, M.C. 1990. Carbon storage forestry. In: *Intergovernmental Panel on Climate Change (IPCC), Response Strategies Working Group (RSWG), Subgroup on Agriculture, Forestry and other Human Activities (AFOS), Proceedings of the Conference on Tropical Forestry Response Options to Global Climate Change, 9–12 January 1990 (UEPA-OPA, PM221)*. Office of Policy Analysis, US Environmental Protection Agency, Washington, DC.

Trexler, M.C. 1991. Estimating tropical biomass futures: a tentative scenario. In: *Proceedings of the Technical Workshop to Explore Options for Global Forestry Management, Bangkok, Thailand 24–30 April 1991*. IIED, London.

Trexler, M.C., Faeth, P.E. and Kramer, J.M. 1989. Forestry as a response to global warming. An analysis of the Guatemala Agroforestry and Carbon Sequestration Project. World Resources Institute, Washington, DC.

Trimble, S.W. and Weirich, F.H. 1987. Reforestation reduces streamflow in the southeastern United States. *Journal of Soil and Water Conservation* Vol 42 No 4.

Tsay, T.B. 1987. Forestry in Japan. *Forestry Chronicle* Vol 63 No 2.

White, M.G. and Eastrop, V.F. 1964. The identity of the iroko (or Mule) gall bug (Hem. Psyllidae). *Entomology Monograph Magazine* 99.

Whitehead, D. 1982. Ecological aspects of natural and plantation forests. *Forestry Abstracts* Vol 43 No 10.

Whitmore, J.L. 1987. Plantation forestry in the tropics of Latin America: a research agenda. *Unasylva* 156 Vol 39 1987/2.

Wicht, C.L. 1967. Forest hydrology research in the South African Republic. In: *Forest Hydrology. Proceedings of an International Symposium on Forest Hydrology.* Pergamon Press, New York.

Wiersum, K.F. and Ketner, P. 1989. Reforestation, a feasible contribution to reducing atmospheric carbon dioxide content? In: Okken, P. A., Swart, R.J. and Zwerver, S. (Eds), *Climate and Energy.* Kluwer Academic Publishers, Dordrecht/Boston/London.

Will, G.M. 1984. Monocultures and site productivity. IUFRO Site and Productivity of Fast Growing Plantations Symposium, Pretoria and Pietermaritzburg, South Africa.

Wilson, W. and Johns, A. 1982. Diversity and abundance of selected animal species in undisturbed forest, selectively logged forest and plantations in East Kalimantan, Indonesia. *Biological Conservation* Vol 24 No 3.

Wormald, T.J. (in preparation). Mixed and pure forest plantations in the tropics and sub-tropics. FAO Forestry Paper. FAO, Rome.

Zobel, B.J. and Davey, C.B. 1977. A conservation miracle. *Alabama Forest Products* 20(5).

Zobel, B J and Talbert, J. 1984. *Applied Forest Tree Improvement.* Wiley, New York.

Zobel, B.J., van Wyk, G. and Stahl, P. 1987. *Growing Exotic Forests.* Wiley-Interscience, New York.